Lecture Notes in Bioinformatics   3737

Subseries of Lecture Notes in Computer Science

T0232834

Corrado Priami   Emanuela Merelli
Pedro Pablo Gonzalez   Andrea Omicini (Eds.)

Transactions on Computational Systems Biology III

 Springer

Series Editors

Sorin Istrail, Brown University, Providence, RI, USA
Pavel Pevzner, University of California, San Diego, CA, USA
Michael Waterman, University of Southern California, Los Angeles, CA, USA

Editor-in-Chief

Corrado Priami
Università di Trento
Dipartimento di Informatica e Telecomunicazioni
Via Sommarive, 14, 38050 Povo (TN), Italy
E-mail: priami@dit.unitn.it

Volume Editors

Emanuela Merelli
Università di Camerino
Via Madonna delle Carceri
62032 Carmerino, Italy
E-mail: emanuela.merelli@unicam.it

Pedro Pablo Gonzalez
Universidad Autonoma Metropolitana
Departamento de Matematicas Aplicadas y Sistemas
Unidad Cuajimalpa, Mexico
E-mail: ppgp@servidor.unam.mx

Andrea Omicini
DEIS, Alma Mater Studiorum, Università di Bologna
Via Venezia 52, 47023 Cesena, Italy
E-mail: andrea.omicini@unibo.it

Library of Congress Control Number: 2005937051

CR Subject Classification (1998): J.3, H.2.8, F.1

ISSN   0302-9743
ISBN-10   3-540-30883-0 Springer Berlin Heidelberg New York
ISBN-13   978-3-540-30883-6 Springer Berlin Heidelberg New York

Springer is a part of Springer Science+Business Media

springeronline.com

© Springer-Verlag Berlin Heidelberg 2005
Printed in Germany

Typesetting: Camera-ready by author, data conversion by Scientific Publishing Services, Chennai, India
Printed on acid-free paper   SPIN: 11599128   06/3142   5 4 3 2 1 0

Preface

In the last few decades, advances in molecular biology and in the research infrastructure in this field has given rise to the "omics" revolution in molecular biology, along with the explosion of databases: from genomics to transcriptomics, proteomics, interactomics, and metabolomics. However, the huge amount of biological information available has left a bottleneck in data processing: information overflow has called for innovative techniques for their visualization, modelling, interpretation and analysis. The many results from the fields of computer science and engineering have then met with biology, leading to new, emerging disciplines such as bioinformatics and systems biology. So, for instance, as the result of application of techniques such as machine learning, self-organizing maps, statistical algorithms, clustering algorithms and multi-agent systems to modern biology, we can actually model and simulate some functions of the cell (e.g., protein interaction, gene expression and gene regulation), make inferences from the molecular biology database, make connections among biological data, and derive useful predictions.

Today, and more generally, two different scenarios characterize the postgenomic era. On the one hand, the huge amount of datasets made available by biological research all over the world mandates for suitable techniques, tools and methods meant at modelling biological processes and analyzing biological sequences. On the other hand, biological systems work as the sources of a wide range of new computational models and paradigms, which are now ready to be applied in the context of computer-based systems.

Since 2001, NETTAB (the International Workshop on Network Tools and Applications in Biology) is the annual event aimed at introducing and discussing the most innovative and promising network tools and applications in biology and bioinformatics. In September 2004, the 4th NETTAB event (NETTAB 2004) was held in the campus of the University of Camerino, in Camerino, Italy. NETTAB 2004 was dedicated to "Models and Metaphors from Biology to Bioinformatics Tools". It brought together a number of innovative contributions from both bioscientists and computer scientists, illustrating their original proposals for addressing many of the open issues in the field of computational biology. Along with an enlightening invited lecture by Luca Cardelli (from Microsoft Research), the presentations and the many lively discussions made the workshop a very stimulating and scientifically profound meeting, which provided the many participants with innovative results and achievements, and also with insights and visions on the future of bioinformatics and computational biology.

This special issue is the result of the workshop. It includes the reviewed and revised versions of a selection of the papers originally presented at the workshop, and also includes a contribution from Luca Cardelli, presenting and elaborating on his invited lecture. In particular, the papers published in this volume cover issues such as:

- data visualization
- protein/RNA structure prediction
- motif finding
- modelling and simulation of protein interaction
- genetic linkage analysis
- notations and models for systems biology

Thanks to the excellent work of the many researchers who contributed to this volume, and also to the patient and competent cooperation of the reviewers, we are confident that this special issue of the LNCS Transactions on Computational Systems Biology will transmit to the reader at least part of the sense of achievement, the dazzling perspectives, and even the enthusiasm that we all felt during NETTAB 2004. A special thanks is then due to the members of the Program Committee of NETTAB 2004, who allowed us, as the Workshop Organizers, to prepare such an exciting scientific program: Russ Altman, Jeffrey Bradshaw, Luca Cardelli, Pierpaolo Degano, Marco Dorigo, David Gilbert, Carole Goble, Anna Ingolfsdottir, Michael Luck, Andrew Martin, Peter McBurney, Corrado Priami, Aviv Regev, Giorgio Valle, and Franco Zambonelli.

Finally, the Guest Editors are very grateful to the Editor-in-Chief, Corrado Priami, for giving them the chance to work on this special issue, and also to the people at Springer, for their patient and careful assistance during all the phases of the editing process.

June 2005   Emanuela Merelli
                                                             Pablo Gonzalez
                                                             Andrea Omicini

LNCS Transactions on
Computational Systems Biology –
Editorial Board

Table of Contents

Computer-Aided DNA Base Calling
from Forward and Reverse Electropherograms

Valerio Freschi and Alessandro Bogliolo

STI - University of Urbino, Urbino, IT-61029, Italy
{freschi, bogliolo}@sti.uniurb.it

Abstract. In order to improve the accuracy of DNA sequencing, forward and reverse experiments are usually performed on the same sample. Base calling is then performed to decode the chromatographic traces (electropherograms) produced by each experiment and the resulting sequences are aligned and compared to obtain a unique consensus sequence representative of the original sample. In case of mismatch, manual editing need to be performed by an experienced biologist looking back at the original traces. In this work we propose computer-aided approaches to base calling from forward and reverse electropherograms aimed at minimizing the need for human intervention during consensus generation. Comparative experimental results are provided to evaluate the effectiveness of the proposed approaches.

1 Introduction

DNA sequencing is an error-prone process composed of two main steps: generation of an *electropherogram* (or *trace*) representative of a DNA sample, and interpretation of the electropherogram in terms of base sequence. The first step entails chemical processing of the DNA sample, electrophoresis and data acquisition [9]; the second step, known as base calling, entails digital signal processing and decoding usually performed by software running on a PC [4,5,6,10]. In order to improve the accuracy and reliability of DNA sequencing, multiple experiments may be independently performed on the same DNA sample. In most cases, forward and reverse experiments are performed by sequencing a DNA segment from the two ends. Bases that appear at the beginning of the forward electropherogram appear (complemented) at the end of the reverse one. Since most of the noise sources are position-dependent (e.g., there is a sizable degradation of the signal-to-noise ratio during each experiment) starting from opposite sides provides valuable information for error compensation. The traditional approach to base calling from opposite traces consists of: i) performing independent base calling on each electropherogram, ii) aligning the corresponding base sequences, and iii) obtaining a consensus sequence by means of comparison and manual editing. The main issue in this process is error propagation: after base calling, wrong bases take part in sequence alignment as if they were correct, although annotated by a confidence level. In case of mismatch, the consensus is manually generated either by comparing the confidence levels of the mismatching bases or by looking back at the original traces.

In this work we explore the feasibility of computer-aided consensus generation. We propose two different approaches. The first approach (called *consensus generation after base calling, CGaBC*) resembles the traditional consensus generation, except for

C. Priami et al. (Eds.): Trans. on Comput. Syst. Biol. III, LNBI 3737, pp. 1–13, 2005.

the fact that automated decisions are taken, in case of mismatch, on the basis of the quality scores assigned by the base caller to forward and reverse sequences. The second approach (called *base calling after trace merging, BCaTM*) performs base calling after error compensation: electropherograms obtained from forward and reverse sequencing experiments are merged in a single averaged electropherogram less sensitive to sequencing errors and noise. Base calling is then performed on the averaged trace directly providing a consensus sequence. The tool flows of the two approaches are shown in Figure 1. Two variants of the second approach are presented differing only for the way the original electropherograms are aligned for merging purposes.

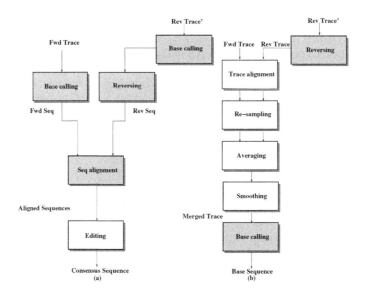

Fig. 1. Tool flow of computer aided base calling from forward and reverse electropherograms: (a) CGaBC. (b) BCaTM. (Rev Trace' denotes the original reverse trace to be reversed and complemented into Rev Trace)

The results presented in this paper show that reliable automated decisions can be taken in most cases of mismatch, significantly reducing the human effort required to generate a consensus sequence. The key issue, however, is how to distinguish between reliable and unreliable automated decisions. A quality score is assigned to this purpose to each base of the consensus sequence. If the quality is above a given threshold, automated decisions can be directly accepted, otherwise they need to be double checked by a human operator. The significance of the quality threshold is discussed in the result section.

1.1   Base Calling

Base calling is the process of determining the base sequence from the electropherogram provided by a DNA automated sequencer. In particular we refer to the DNA sequencing process known as Sanger's method [9]. Four reactions of extension from initial primers of a given template generate an entire set of nested sub-fragments in which the last base of every fragment is marked with 4 different types of fluorescent markers (one for each type of base). Fragments are then sorted by length by means of capillary electrophoresis and detected by 4 optical sensors working at disjoint wavelengths in order to distinguish the emissions of the 4 markers. The result of a sequencing experiment is an electropherogram that is a 4-component time series made of the samples of the emissions measured by the 4 optical sensors. In principle, the DNA sequence can be obtained from the electropherogram by associating each dominant peak with the corresponding base type and by preserving the order of the peaks. However, electropherograms are affected by several non-idealities (random noise of the measuring equipment, cross-talk due to the spectral overlapping between fluorescent markers, sequence-dependent mobility, ...) that require a pre-processing step before decoding. Since the non-idealities depend on the sequencing strategy and on the sequencer, pre-processing (consisting of multi-component analysis, mobility shift compensation and noise filtering) is usually performed by software tools distributed with sequencing machines [1]. The original electropherogram is usually called *raw data*, while we call *filtered data* the result of pre-processing. In the following we will always refer to filtered data, representing the filtered electropherogram (hereafter simply called *electropherogram*, or *trace*, for the sake of simplicity) as a sequence of 4-tuples. The k-th 4-tuple (A_k, C_k, G_k, T_k) represents the emission of the 4 markers at the k-th sampling instant, uniquely associated with a position in the DNA sample. In this paper we address the problem of base calling implicitly referring to the actual decoding step, that takes in input the (filtered) electropherogram and provides a base sequence. Base calling is still a difficult and error-prone task, for which several algorithms have been proposed [4,6,10]. The result they provide can be affected by different types of errors and uncertainties: *mismatches* (wrong base types at given positions), *insertions* (bases artificially inserted by the base caller), *deletions* (missed bases), *unknowns* (unrecognized bases, denoted by N). The accuracy of a base caller can be measured both in terms of number of N in the sequence, and in terms of number of errors (mismatches, deletions and insertions) with respect to the actual consensus sequence. The accuracy obtained by different base callers starting from the same electropherograms provides a fair comparison between the algorithms. On the other hand, base callers usually provide estimates of the quality of the electropherograms they start from [3]. A quality value is associated with each called base, representing the correctness probability: the higher the quality the lower the error probability. Since our approach generates a synthetic electropherogram to be processed by a base caller, in the result section we also compare quality distributions to show the effectiveness of the proposed technique.

1.2   Sequence Alignment

Sequence comparison and alignment are critical tasks in many genomic and proteomic applications. The best alignment between two sequences F and R is the alignment that minimizes the effort required to transform F in R (or vice versa). In general, each edit operation (base substitution, base deletion, base insertion) is assigned with a cost, while each matching is assigned with a reward. Scores (costs and rewards) are empirically assigned depending on the application. The score of a given alignment between F and R is computed as the difference between the sum of the rewards associated with the pairwise matches involved in the alignment, and the sum of the edit operations required to map F onto R. The best alignment has the maximum similarity score, that is usually taken as similarity metric. The basic dynamic programming algorithm for computing the similarity between a sequence F of M characters and a sequence R of N characters was proposed by Needleman and Wunsch in 1970 [8], and will be hereafter denoted by NW-algorithm. It makes use of a score matrix D of $M + 1$ rows and $N + 1$ columns, numbered starting from 0. The value stored in position (i, j) is the similarity score between the first i characters of F and the first j characters of R, that can be incrementally obtained from $D(i - 1, j)$, $D(i - 1, j - 1)$ and $D(i, j - 1)$:

$$D(i, j) = max \begin{cases} D(i - 1, j - 1) + S_{sub}(F(i), R(j)) \\ D(i - 1, j) + S_{del} \\ D(i, j - 1) + S_{ins} \end{cases} \tag{1}$$

S_{ins} and S_{del} are the scores assigned with each insertion and deletion, while S_{sub} represents either the cost of a mismatch or the reward associated with a match, depending on the symbols associated with the row and column of the current element. In symbols, $S_{sub}(F(i), R(j)) = S_{mismatch}$ if $F(i) \neq R(j)$, $S_{sub}(F(i), R(j)) = S_{match}$ if $F(i) = R(j)$.

2   Consensus Generation after Base Calling (CGaBC)

When forward and reverse electropherograms are available, the traditional approach to determine the unknown DNA sequence consists of: i) independently performing base calling on the two traces in order to obtain forward and reverse sequences, ii) aligning the two sequences and iii) performing a minimum number of (manual) editing steps to obtain a consensus sequence. The flow is schematically shown in Figure 1.a, where the reverse trace is assumed to be reversed and complemented by the processing block labeled Reverse. Notice that complementation can be performed either at the trace level or at the sequence level (i.e., after base calling). In Fig. 1.a the reverse trace is reversed and complemented after base calling.

The results of the two experiments are combined only once they have been independently decoded, without taking advantage of the availability of two chromatographic traces to reduce decoding uncertainties. Once base-calling errors have been made on each sequence, wrong bases are hardly distinguishable from correct ones and they take part in alignment and consensus. On the other hand, most base callers assign with each base a quality (i.e., confidence) value (representing the correctness probability) computed on the basis of the readability of the trace it comes from.

In a single sequencing experiment, base qualities are traditionally used to point out unreliable calls to be manually checked. When generating a consensus from forward and reverse sequences, quality values are compared and combined. Comparison is used to decide, in case of mismatch, for the base with higher value. Combination is used to assign quality values to the bases of the consensus sequence.

The proper usage of base qualities has a great impact on the accuracy (measured in terms of errors) and completeness (measured in terms of undecidable bases) of the consensus sequence. However, there are no standard methodologies for comparing and combining quality values.

The CGaBC approach presented in this paper produces an *aggressive consensus* by taking always automated decisions based on base qualities: in case of a mismatch, the base with higher quality is always assigned to the consensus. Since alignment may give rise to gaps, quality values need also to be assigned to gaps. This is done by averaging the qualities of the preceding and following calls.

Qualities are assigned to the bases of the consensus sequence by adding or subtracting the qualities of the aligned bases in case of match or mismatch, respectively [7]. Quality composition, although artificial, is consistent with the probabilistic nature of quality values, defined as $q = -10log_{10}(p)$, where p is the estimated error probability for the base call [5].

In some cases, however, wrong bases may have quality values greater than correct ones, making it hard to take automated correct decisions. The overlapping of the quality distributions of wrong and correct bases is the main problem of this approach.

A quality threshold can be applied to the consensus sequence to point out bases with a low confidence level. Such bases need to be validated by an experienced operator, possibly looking back at the original traces.

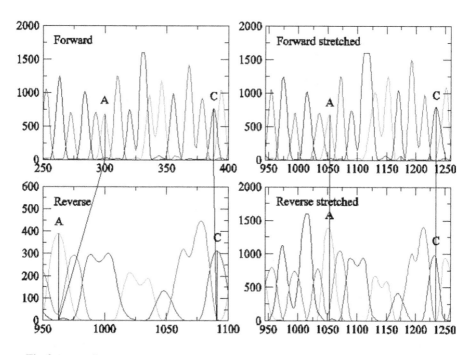

Fig. 2. Trace alignment issues (left) and sample re-positioning on a common x-axis (right)

3   Base Calling after Trace Merging (BCaTM)

The approach is illustrated in Figure 1b. We first obtain an average trace by combining the two experiments available for the given DNA, then we perform base calling on the averaged trace directly obtaining the consensus sequence. The rationale behind this approach is two-fold. First, electropherograms are much more informative than the corresponding base sequences, so that their comparison provides more opportunities for noise filtering and error correction. Second, each electropherogram is the result of a complex measurement experiment affected by random errors. Since the average of independent measurements has a lower standard error, the averaged electropherogram has improved quality with respect to the original ones.

Averaging independent electropherograms is not a trivial task, since they usually have different starting points, different number of samples, different base spacing and different distortions, as shown in the left-most graphs of Fig. 2. In order to compute the point-wise average of the traces, we need first to re-align the traces so that samples belonging to the same peak (i.e., representing the same base) are in the same position, as shown in the right-most graphs of Fig. 2. By doing this, we are then able to process *homologous* samples, that is to say samples arranged according to the fact that in the same position on the x-axis we expect to find values representing the same point of the DNA sample. A similar approach was used by Bonfield et al. [2] to address a different problem: comparing two electropherograms to find point mutations. However, the authors didn't discuss the issues involved in trace alignment and point-wise manipulation.

We propose two different procedures for performing trace alignment. The first is based on the maximization of the correlation between the four time series, using a dynamic programming algorithm derived form the NW-algorithm. The second makes use of a preliminary base calling step to identify base positions within the trace to be used to drive trace alignment. The overall procedures (respectively denoted as *sample-driven alignment* and *base-driven alignment*) are described in the next sections, assuming that forward and reverse traces are available and that the reverse trace has already been reversed and complemented. All procedures are outlined in Fig. 3.

After alignment, forward and reverse traces are re-sampled using a common sampling step and their sample-wise average is computed to obtain the averaged electropherogram. Local smoothing is then performed to remove small artificial peaks possibly introduced by the above steps. Base calling is finally performed on the averaged electropherogram directly providing the consensus sequence. The entire process is outlined in Fig. 1.b.

3.1   Sample-Driven (SD) Trace Alignment

Sample-driven trace alignment aims at maximizing the correlation between the 4-component time series that constitute forward and reverse electropherograms. Aligned electropherograms are nothing but the original electropherograms with labels associated with samples to denote pairwise alignment. Homologous samples share the same label. The score associated with an alignment is the sample-wise correlation computed according to the alignment.

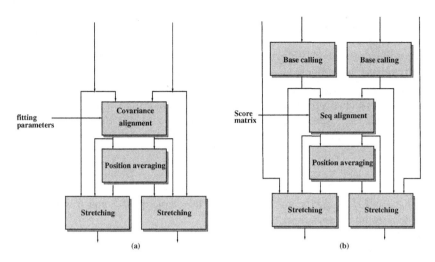

Fig. 3. a) Sample-driven alignment procedure. b) Base-driven alignment procedure.

The difference between pairwise correlation of electropherograms and pairwise correlation of standard time series is twofold. First, electropherograms are 4-component time series. The correlation between two electropherograms is the sum of the pairwise correlations between their homologous components. Second, due to the intrinsic nature of electrophoretic runs, electropherograms might need to be not only shifted, but also stretched with respect to each other in order to obtain a meaningful point-wise alignment. Stretching can be obtained by means of gap insertion at specific positions of one of the two electropherograms under alignment.

Despite the above-mentioned peculiarities, the correlation between electropherograms retains the additive property of the standard correlation. Hence, the alignment corresponding to the maximum correlation can be incrementally determined by means of dynamic programming techniques. In the next subsection we outline a modified version of the NW algorithm that handles electropherograms maximizing their correlation.

Dynamic Programming Alignment Algorithm. For the sake of simplicity, we sketch the NW modified algorithm considering single-component traces. We will then generalize to the 4-component case. As previously introduced in section 1.2, the NW algorithm incrementally computes the optimal path through the dynamic-programming matrix (DP matrix) according to a specific optimality criterion. At each step a new entry (say, i,j) of the matrix is computed as the maximum score achieved by means of one of the three possible moves leading to that position: a diagonal move that adds a replacement score (that is a reward for the alignment of the ith element of sequence F with the jth element of sequence R) to the value stored in entry $(i-1, j-1)$; a vertical move that adds a deletion cost to the value stored in entry $(i-1, j)$; and a horizontal move that adds an insertion cost to the value stored in entry $(i, j-1)$. As far as alignment is concerned, insertions and deletions are symmetric moves: deleting an element from sequence F has the same effect of adding an element in sequence R. Although both insertions and deletions are needed to stretch the two sequences in order to achieve the best alignment between them, the two operations are nothing but gap insertions in one of the two sequences.

According to the above observations, we outline the modified NW algorithm referring to two basic operations: alignment between existing samples of the two electropherograms (corresponding to a diagonal move in DP matrix) and insertion of a gap in one of the two electropherograms (corresponding to vertical or horizontal moves in the DP matrix). The score to be assigned to the alignment between existing samples of the two electropherograms (diagonal move) is computed as the correlation between the samples.

$$S_{diag}(i,j) = \frac{(F(i) - F_{avg})(R(j) - R_{avg})}{\sigma_F \sigma_R}$$

where F_{avg} and R_{avg} are the average values of the elements of F and R, while σ_F and σ_R are their standard deviations.

In order to assign a score to a gap we refer to the role the gap will play in the final alignment. After alignment, the two aligned electropherograms need to be processed in order to fill all the gaps possibly inserted by the NW algorithm. Synthetic samples need to be created to this purpose and added at proper positions. Such synthetic samples are introduced by interpolating the existing samples on both sides of a gap. In this perspective, the score to be assigned to a gap insertion in one of the two electropherograms (vertical or horizontal moves) is computed as the correlation between the synthetic sample (generated by interpolation) to be added to bridge the gap and the original sample of the other trace aligned with the gap. The score assigned with an horizontal move leading to entry (i, j), corresponding to a gap insertion in the forward trace F, will be:

$$S_{hor}(i,j) = \frac{(\frac{F(i)+F(i+1)}{2} - F_{avg})(R(j) - R_{avg})}{\sigma_F \sigma_R}$$

while the score assigned with a vertical move will be:

$$S_{ver}(i,j) = \frac{(F(i) - F_{avg})(\frac{R(j)+R(j+1)}{2} - R_{avg})}{\sigma_F \sigma_R}$$

If we deal with 4-component time series rather than with single-component traces, we can extend the algorithm by maximizing the sum of the four correlations:

$$S_{diag}(i,j) = \sum_{h \in \{A,C,G,T\}} \frac{(F^{(h)}(i) - F^{(h)}_{avg})(R^{(h)}(j) - R^{(h)}_{avg})}{\sigma_F^{(h)} \sigma_R^{(h)}}$$

$$S_{hor}(i,j) = \sum_{h \in \{A,C,G,T\}} \frac{(\frac{F^{(h)}(i)+F^{(h)}(i+1)}{2} - F^{(h)}_{avg})(R^{(h)}(j) - R^{(h)}_{avg})}{\sigma_F^{(h)} \sigma_R^{(h)}} \quad (2)$$

$$S_{ver}(i,j) = \sum_{h \in \{A,C,G,T\}} \frac{(F^{(h)}(i) - F^{(h)}_{avg})(\frac{R^{(h)}(j)+R^{(h)}(j+1)}{2} - R^{(h)}_{avg})}{\sigma_F^{(h)} \sigma_R^{(h)}}$$

where index h spans the four components A, C, G and T.

Although equation 2 leads to a mathematically sound computation of the correlation between forward and reverse traces, the alignment corresponding to the maximum correlation is not robust enough with respect to base calling. This is due to the fact that all samples have the same weight in the correlation, while different weights are implicitly assigned to the samples by the base caller. In fact, the attribution of bases is strongly dependent on the identification of local maxima, dominant point values and slopes. Neglecting this properties during the alignment of the traces leads to poor results.

In order to capture all the features of the original electropherograms that take part in base calling, the score functions of equation 2 need to be modified as follows.

First, the dominant component of each sample is weighted twice when computing the incremental correlation. For instance, if the ith sample of trace F has components $F^{(A)}(i) = 23, F^{(C)}(i) = 102, F^{(G)}(i) = 15, F^{(T)}(i) = 41$, the correlation between $F^{(C)}(i)$ and $R^{(C)}(j)$ is added twice to the local score.

Second, the derivatives of the time series are computed and their correlations added to the local scores. In particular, both first and second order derivatives are taken into account to obtain a good alignment of local maxima.

The overall score assigned with each move is then computed as the weighted sum of three terms: the correlation between the 4-component time series (modified to reward dominant components), the correlation between the 4-components derivatives and the correlation between the 4-component second-order derivatives. The weights of the three terms are fitting parameters that need to be tuned to obtain good results on a set of known benchmarks. For our experiments we used weights 1, 2 and 3 for correlations between time series, first-order derivatives and second-order derivatives.

Post-processing of Aligned Traces. The outputs of trace alignment are two sequences of samples with gaps inserted in both traces whenever required to obtain the maximum-score alignment. In order to make it possible to perform a sample-wise average of the aligned traces, we need to replace the gaps possibly inserted during alignment with artificial (yet meaningful) samples. To this purpose we replace gaps with 4-component samples obtained by means of linear interpolation. Consider, for instance, the situation of Figure 4, where a few samples of two aligned traces are shown. A double gap has been inserted in Trace F by the alignment algorithm. We fill the gap in position 103, with a component-wise linear interpolation between samples 102 and 105 of the same trace. For instance, the new value of $F^{(C)}(103)$ will be:

$$F^{(C)}(103) = F^{(C)}(102) + (F^{(C)}(105) - F^{(C)}(102)) \cdot \frac{103 - 102}{105 - 102} = 110 - 3 \cdot \frac{1}{3} = 109$$

3.2   Base-Driven (BD) Trace Alignment

Base-driven trace alignment is outlined in Figure 3b. First of all we perform independent base calling on the original traces annotating the position (i.e., the point in the trace) of each base. Base sequences are then aligned by means of the NW-algorithm, possibly inserting gaps (represented by character '-') between original bases. Since mismatches between forward and reverse traces obtained by sequencing the same sample can be caused either by random noise or by decoding errors, we assign the same cost (namely, -1) to each edit operation performed by the NW-algorithm (insertion, deletion,

Trace F	A:	...	123	–	–	131	...
	C:	...	110	–	–	107	...
	G:	...	12	–	–	12	...
	T:	...	1	–	–	9	...
Common position		101	102	103	104	105	106
Trace R	A:	...	180	170	166	160	...
	C:	...	40	55	67	89	...
	G:	...	5	7	8	8	...
	T:	...	10	8	9	11	...

Fig. 4. Example of aligned traces with a double gap inserted in Trace F

substitution), while we assign a positive reward (+1) to each match. Sequence alignment is then used to drive trace alignment as described below.

From Sequence Alignment to Trace Alignment. The outputs of the NW-algorithm are aligned base sequences with annotated positions in the original traces. What we need to do next is to handle gaps and to use base alignment for inducing the alignment of the corresponding traces. We associate a virtual position to each missing base (i.e., to each gap) assuming that missing bases are equally spaced from the preceding and following bases. All bases are then re-numbered according to the alignment, taking into account the presence of gaps. Aligned bases on the two traces are associated with the same number. Bases belonging to the two traces are then re-positioned on a common x axis, so that homologous (aligned) bases have the same position. The new position (x_k) of base k is computed as the average of the positions of the k-th bases on the two original traces. Then, the original traces are shrunk in order to adapt them to the common x axis. This entails re-positioning not only trace samples associated with base calls, but all the original samples in between. After shrinking, the peaks associated with the k-th base on the two traces are in the same position x_k, as shown in the right-most graphs of Fig. 2.

The base-driven approach described in this section is less intuitive than the sample-driven approach described in Section 3.1. In fact, it relies on base calling to drive trace composition that, in its turn, is used to improve the accuracy of base calling. In practice, preliminary base calling is used to filter out from the original electropherograms the noisy details that may impair the robustness of trace alignment. On the other hand, the bases called during the preliminary step affect only the alignment, while they do not affect trace merging. All the samples of the original electropherograms, possibly stretched/shrunk according to the alignment, are used to generate the merged trace. As a result, the merged trace is much more informative than a consensus sequence derived from the alignment, and it usually has improved quality with respect to the original electropherograms thanks to error compensation.

4   Results and Conclusions

We tested all approaches on a set of known DNA samples. The experiments reported are not sufficient for a thorough accuracy assessment of the proposed approaches. Rather, they must be regarded as case studies and will be reported and discussed in detail in order to point out the strengths and weaknesses of the methodologies presented in this paper.

For each sample, forward and reverse electropherograms were obtained using an *ABI PRISM 310 Genetic Analyzer* [1]. *Phred* software [4] was used for base calling, while procedures for consensus generation, trace alignment and averaging were implemented in C and run on a PC under *Linux*. For each sample we generated three sets of results by applying consensus generation after base calling (CGaBC), base calling after sample-driven trace merging (SD BCaTM) and base calling after base-driven trace merging (BD BCaTM). The accuracy of each solution was evaluated by pairwise alignment against the known actual sequence.

Experimental results are reported in table 1. The first two column reported name of the experiment and the length (i.e., number of bases, namely L) of the overlapping region of forward and reverse traces. Columns 3-5 and 6-8 refer to the sequences obtained by means of base calling from the original forward and reverse traces, respectively. The accuracy of each sequence is shown in terms of *number of errors* (E) (computed including unrecognized bases and calling errors), *maximum quality value assigned to a wrong call (MQw)* and *minimum quality value assigned to a correct call (mQc)*. Since quality values are mainly used by biologists to discriminate between reliable and unreliable calls, ideally it should be $MQw < mQc$. Unfortunately this is often not the case. When $MQw > mQc$, the quality distributions of wrong and correct bases overlap, making it hard to set an acceptance threshold. Column labeled X reports the number of mismatches between forward and reverse sequences. This provides a measure of the number of cases requiring non-trivial decisions that may require human intervention. Finally, columns 10-12, 13-15 and 16-18 report the values of E, MQw and mQc for the consensus sequences generated by the three automated approaches.

Benchmarks are ordered in both tables based on the average qualities assigned by Phred to the Forward and Reverse calls. The first 2 samples have average quality lower than 20, samples 3 and 4 have average quality lower than 30, all other samples have

Table 1. Experimental results

Name	L	Forward			Reverse			X	CGaBC (Consensus)			SD BCaTM (Sample-driven)			BD BCaTM (Base-driven)		
		E	MQw	mQc	E	MQw	mQc		E	MQw	mQc	E	MQw	mQc	E	MQw	mQc
3B	47	4	15	6	2	15	9	4	3	33	14	19	20	6	0	-	8
9B	200	2	25	7	35	15	6	37	5	28	0	7	20	6	4	11	6
5B	191	4	10	6	11	12	6	12	4	18	0	3	7	6	4	9	6
5C	182	1	9	7	6	14	8	5	0	-	2	4	12	4	1	9	9
8C	216	8	12	4	8	10	7	15	0	-	17	3	15	4	1	11	11
10B	166	1	7	7	4	17	8	5	0	-	29	1	19	10	0	-	10
10A	203	0	-	8	4	12	7	4	0	-	22	1	7	6	0	-	9
16A	228	5	9	4	10	15	4	14	0	-	10	0	-	9	6	11	6
1AT	559	2	13	8	5	17	7	7	3	31	1	0	-	9	0	-	9
2AT	562	2	9	4	6	18	4	8	1	35	9	0	-	9	0	-	11
Avg	255	2.9	12.1	6.1	9.1	14.5	6.6	11.1	1.6	29.0	10.4	3.8	14.3	6.9	1.6	10.2	8.5

average quality above 30. All results are summarized in the last row, that reports column averages.

First of all we observe that all proposed approaches produces a number of errors much lower than the number of X's, meaning that correct automated decisions are taken in most cases.

The weakest approach seems to be SD BCaTM, whose effectiveness is strongly dependent on the quality of the original traces: the number of errors made on lowest-quality trace is much higher than the number of X's (meaning that it is counterproductive) while it becomes lower when starting from good traces (see samples 1AT and 2AT).

CGaBC and BD BCaTM seem much more robust: the number of errors they made is always much lower than the number of X's and their accuracy is almost independent of the quality of the original traces.

It is also worth remarking that the quality values assigned by Phred to sequences directly called from merged traces are much more informative than the combined values assigned by CGaBC to consensus sequences. This is shown by the difference between mQc and MQw, that is (on average) -1.7 for the base-driven approach, while it is -18.6 for the aggressive consensus, denoting a much smaller overlapping between the qualities assigned to wrong and correct base calls.

The effectiveness of quality thresholds used to discriminate between reliable and unreliable base calls is further analyzed in Fig. 5. The percentage of *false negatives* (i.e., correct base calls regarded as unreliable) and *false positives* (i.e., wrong base calls

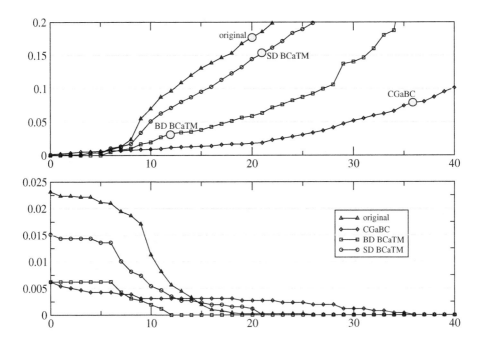

Fig. 5. Comparison of quality values of Original base calls, CGaBC, SD BCaTM and BD BCaTM

regarded as reliable) are plotted as functions of the quality threshold for: the sequences called from original forward and reverse traces (*original*), the aggressive consensus sequences (*CGaBC*), and the sequences called from sample-driven (*SD BCaTM*) and base-driven (*BD BCaTM*) merged traces. Statistics are computed over all the case studies of table 1.

Given a reliability constraint (i.e., an acceptable false-positive probability), the corresponding quality threshold can be determined on the bottom graph. The corresponding percentage of false negatives can be read from the upper graph. The percentage of false negatives is a measure of the number of base calls that need to be checked by a human operator in order to obtain the required reliability. Large circular marks on the top graph denote the percentage of false negatives corresponding to quality thresholds that reduce to 0 the risk of false positives. The lower the value, the better. Interestingly, the percentage is of 4.7% for BD BCaTM, while it is around 17.5% for the original traces, around 16% for SD BCaTM and around 8.05% for the aggressive consensus. This is a measure of the improved quality of the merged traces.

In conclusion, we have presented and compared automated approaches to base calling from forward and reverse electropherograms. Experimental results show that the proposed techniques provide significant information about hard-to-decide calls that may be used to reduce the human effort required to construct a reliable sequence. In particular, the so called BD BCaTM and CGaBC approaches produced the best results, making on average 1.6 errors per sequence against an average number of 11.1 mismatches that would require human decisions. Moreover, the improved quality of averaged electropherograms makes it easier to discriminate between correct and incorrect base calls.

References

1. Applied Biosystems: ABI PRISM 310 Genetic Analyzer: user guide, PE Applied Biosystems (2002).
2. J.K. Bonfield, C. Rada, R. Staden: Automated detection of point mutations using fluorescent sequence trace subtraction. Nucleic Acid Res. **26** (1998) 3404-3409.
3. R. Durbin, S. Dear: Base Qualities Help Sequencing Software. Genome Res. **8** (1998) 161-162.
4. B. Ewing, L. Hillier, M.C. Wendl, P. Green: Base-calling of automated sequencer traces using Phred I. Accuracy assessment, Genome Res. **8** (1998) 175-185.
5. B. Ewing, L. Hillier, M.C. Wendl, P. Green: Base-calling of automated sequencer traces using Phred II. Error probabilities, Genome Res. **8** (1998) 186-194.
6. M.C. Giddings, J. Severin, M. Westphall, J. Wu, L.M. Smith: A Software System for Data Analysis in Automated DNA Sequencing, Genome Res. **8** (1998) 644-665.
7. X. Huang, A. Madan: CAP3: A DNA Sequence Assembly Program, Genome Res. **9** (1999) 868-877.
8. S.B. Needleman, C.D. Wunsch: A general method applicable to the search for similarities in the amino acid sequences of two proteins, J.Mol.Biol **48** (1970) 443-453.
9. F. Sanger, S. Nickler, A.R. Coulson,A.R.: DNA sequencing with chain terminating inhibitors, in Proc. Natl. Acad. Sci. **74** (1977) 5463-5467.
10. D. Walther, G. Bartha, M. Morris: Basecalling with LifeTrace. Genome Res. **11** (2001) 875-888.

A Multi-agent System for Protein Secondary Structure Prediction

Giuliano Armano[1], Gianmaria Mancosu[2], Alessandro Orro[1], and Eloisa Vargiu[1]

[1] University of Cagliari, Piazza d'Armi, I-09123, Cagliari, Italy
{armano, orro, vargiu}@diee.unica.it
http://iasc.diee.unica.it
[2] Shardna Life Sciences, Piazza Deffenu 4, I-09121 Cagliari, Italy
mancosu@shardna.it

Abstract. In this paper, we illustrate a system aimed at predicting protein secondary structures. Our proposal falls in the category of multiple experts, a machine learning technique that –under the assumption of absent or negative correlation in experts' errors– may outperform monolithic classifier systems. The prediction activity results from the interaction of a population of experts, each integrating genetic and neural technologies. Roughly speaking, an expert of this kind embodies a genetic classifier designed to control the activation of a feedforward artificial neural network. Genetic and neural components (i.e., guard and embedded predictor, respectively) are devoted to perform different tasks and are supplied with different information: Each guard is aimed at (soft-) partitioning the input space, insomuch assuring both the diversity and the specialization of the corresponding embedded predictor, which in turn is devoted to perform the actual prediction. Guards deal with inputs that encode information strictly related with relevant domain knowledge, whereas embedded predictors process other relevant inputs, each consisting of a limited window of residues. To investigate the performance of the proposed approach, a system has been implemented and tested on the RS126 set of proteins. Experimental results point to the potential of the approach.

1 Introduction

Due to the strict relation between protein function and structure, the prediction of protein 3D-structure has become in the last years one of the most important tasks in bioinformatics. In fact, notwithstanding the increase of experimental data on protein structures available in public databases, the gap between known sequences (165,000 entries in Swiss-Prot [5] on Dec 2004) and known tertiary structures (28,000 entries in PDB [8] on Dec 2004) is constantly increasing. The need for automatic methods has brought the development of several prediction and modelling tools, but despite the increase of accuracy a general methodology to solve the problem has not been yet devised. Building complete protein tertiary structure is still not a tractable task, and most methodologies concentrate on the simplified task of predicting their secondary structure. In fact, the

C. Priami et al. (Eds.): Trans. on Comput. Syst. Biol. III, LNBI 3737, pp. 14–32, 2005.

knowledge of secondary structure is a useful starting point for further investigating the problem of finding protein tertiary structures and functionalities. In this paper we concentrate on the problem of predicting secondary structures using a multiple expert system rooted in two powerful soft-computing techniques, i.e. genetic and neural. In Section 2 some relevant work is briefly recalled. Section 3 introduces the proposed multiple expert architecture focusing on the most relevant customizations adopted for dealing with the task of secondary structures prediction. Section 4 illustrates experimental results. Section 5 draws conclusions and future work.

2   Related Work

In this section, some relevant related work is briefly recalled, according to both an applicative and a technological perspective. The former is mainly focused on the task of secondary structure prediction, whereas the latter concerns the subfield of multiple experts, which the proposed system stems from.

2.1   Protein Structure Prediction

Tertiary Structure Prediction. The problem of protein tertiary structure prediction is very complex, as the underlying process involves biological, chemical, and physical interactions. Let us briefly recall three main approaches: (a) comparative modeling, (b) ab-initio methods, and (c) fold recognition methods.

Comparative modeling methods are based on the assumption that proteins with similar sequences fold into similar structures and usually have similar functions [15]; thus, prediction can be performed by comparing the primary structure of a target sequence against a set of protein with known structures. A procedure aimed at building comparative models usually follows three steps [58] [60]: (1) identify proteins with known 3D structures that are related to the target sequence, (2) align proteins with the target sequence and consider each alignment with high score as a template, (3) build a 3D model of the target sequence based on alignment with templates. Several algorithms have been devised to calculate the model corresponding to a given alignment. In particular, let us recall the assembling of rigid bodies [27], segment matching [66] and spatial restraints [28].

Assuming that the native structure of a protein in thermodynamic equilibrium corresponds to its minimum energy state, ab-initio methods build a model for predicting structures by minimizing an energy function. An ab-initio method is composed by: (1) a biological model to represent interactions between different parts of proteins, (2) a function representing the free energy and (3) an optimization algorithm to find the best configuration. Accurate models built at atomic level [10] are feasible only for short sequences; hence, simplified models are needed –which include using either the residue level [45] or a lattice model for representing proteins [62]. Energy functions include atom-based potentials [56], statistical potentials of mean force [67] and simplified potentials based on chemical intuition [33]. Finally, optimization strategies include genetic algorithm [21], Monte Carlo simulation [22], and molecular dynamics simulations [44].

Fold recognition methods start with the observation that proteins encode only a relatively small number of distinct structure topologies [16], [9] and that the estimated probability of finding a similar structure while searching for a new protein in the PDB is about 70% [48]. Thus, a protein structure can be conjectured by comparing the target sequence with those found in a suitable database of structural templates –producing a list of scores. Then, the structure of the target sequence is assigned according to the best score found. Several comparing strategies are defined in the literature. In 3D-1D fold recognition methods [11] the known tertiary structures in the database are encoded into strings representing structural informations, like secondary structure and solvent accessibility. Then, these strings are aligned with the 1D string derived in the same way by the query sequence. A number of variations on this approach exist. For instance, instead of the environment description of the 3D-1D profile, energy potentials can be used [38]. Other comparing strategies include two-level dynamic programming [65], frozen approximation [23], branch and bound [43], Monte Carlo algorithms [46] and heuristic search based on predicted secondary structures [57].

Secondary Structure Prediction. Difficulties in predicting protein structure are mainly due to the complex interactions between different parts of the same protein, on the one hand, and between the protein and the surrounding environment, on the other hand. Actually, some conformational structures are mainly determined by local interactions between near residues, whereas others are due to distant interactions in the same protein. Moreover, notwithstanding the fact that primary sequences are believed to contain all information necessary to determine the corresponding structure [3], recent studies demostrate that many proteins fold into their proper three-dimensional structure with the help of molecular chaperones that act as catalysts [25], [31]. The problem of identifying protein structures can be simplified by considering only their secondary structure; i.e. a linear labeling representing the conformation to which each residue belongs. Thus, secondary structure is an abstract view of amino acid chains, in which each residue is mapped into a secondary alphabet usually composed by three symbols: alpha-helix (α), beta-sheet (β), and random-coil (c). Assessing the secondary structure can help in building the complete protein structure, and can be useful information for making hypoteses on the protein functionality also in absence of information about the tertiary structure. In fact, very often, active sites are associated with a particular conformation or combination (motifs) of secondary structures conserved during the evolution.

There are a variety of secondary structure prediction methods proposed in the literature. Early prediction methods were based on statistics headed at evaluating, for each amino acid, the likelihood of belonging to a given secondary structure [17]. The main drawback of these techniques is that, typically, no contextual information is taken into account, whereas nowadays it is well known that secondary structures are determined by chemical bonds that hold between spatially-close residues. A second generation of methods exhibits better performance by exploiting protein databases, as well as statistic information about amino acid subsequences. In this case, prediction can be viewed as a machine-learning problem, aimed at inferring a shallow model able to correctly classify each residue, usually taking into account a limited window of aminoacids (e.g., 11 continuous residues) –centered around the one being predicted. Several methods exist in this category, which may be classified according to (i) the underlying approach, e.g., statistical information [54], graph-theory [47], multivariate statistics [41], (ii) the kind of information actually taken into account, e.g., physico-chemical properties [50], sequence patterns [64], and (iii) the adopted technique, e.g., k-Nearest Neighbors [59], Artificial Neural Networks (ANNs) [30] (without going into further details, let us only stress that ANNs are the most widely acknowledged technique).

The most significant innovation introduced in prediction systems was the exploitation of long-range and evolutionary information contained in multiple alignments. It is well known, in fact, that even a single variation in a sequence may dramatically compromise its functionality. To figure out which substitutions possibly affect functionality, sequences that belong to the same family can be aligned, with the goal of highlighting regions that preserve a given functionality. The underlying motivation is that active regions of homologous sequences will typically adopt the same local structure, irrespective of local sequence variations. PHD [55] is one of the first ANN-based methods that make use of evolutionary information to perform secondary structure prediction. In particular, after searching similar sequences using BLASTP [1], ClustalW [32] is invoked to identify which residues can actually be substituted without compromising the functionality of the target sequence. To predict the secondary structure of the target sequence, the multiple alignment produced by ClustalW is given as input to a multi layer ANN. The first layer outputs a sequence-to-structure prediction which is sent to a further ANN layer that performs a structure-to-structure prediction aimed at refining it.

Further improvements are obtained with both more accurate multiple alignment strategies and more powerful neural network structures. For instance, PSI-PRED [2] exploits the position-specific scoring matrix (called "profile") built during a preprocessing performed by PSI-BLAST [39]. This approach outperforms PHD thanks to the PSI-BLAST ability of detecting distant homologies. In more recent work [6] [7], Recurrent ANNs (RANNs) are exploited to capture long-range interactions. The actual system that embodies such capabilities, i.e., SSPRO [51], is characterized by: (a) PSI-BLAST profiles for encoding inputs, (ii) Bidirectional RANNs, and (iii) a predictor based on ensembles of RANNs.

2.2   Multiple Experts

Divide-and-conquer is the one of the most popular strategies aimed at recursively partitioning the input space until regions of roughly constant class membership are obtained. Several machine learning approaches e.g., Decision Lists (DL) [53], [19], Decision Trees (DT) [49], Counterfactuals (CFs) [70], Classification And Regression Trees (CART) [12] apply this strategy to control the complexity of the search, thus yielding a monolithic solution of the problem. Nevertheless, a different interpretation can be given, in which the partitioning procedure is considered as a tool for generating multiple experts. Although with a different focus, this multiple experts' perspective has been adopted by the evolutionary-computation and by the connectionist communities. In the former, the focus was on devising suitable architectures and techniques able to enforce an adaptive behavior on a population of individuals, e.g., Genetic Algorithms (GAs) [34], [26], Learning Classifier Systems (LCSs) [35], [36], and eXtended Classifier Systems (XCSs) [72]. In the latter, the focus was mainly on training techniques and output combination mechanisms; in particular, let us recall Jordan's Mixtures of Experts [37], [40] and Weigend's Gated Experts [71]. Further investigations are focused on the behavior of a population of multiple (heterogeneous) experts with respect to a single expert. Theoretical studies and empirical results, rooted in the computational and/or statistical learning theory (see, for instance, [68] and [69]), have shown that the overall performance can be significatively improved by adopting an approach based on multiple experts. Relevant studies in this subfield include Artificial Neural Network (ANN) ensembles [42], [13] and DT ensembles [24], [61].

3   Predicting Secondary Structures Using NXCS Experts

This section introduces the multiple expert architecture devised to tackle the the task of predicting protein secondary structure, which is a customization of the generic NXCS architecture (see, for instance, [4]). NXCS stands for Neural XCS, as the architecture integrates the XCS and ANN technologies. The architecture is illustrated from two separate perspectives, i.e., (i) micro- and (ii) macro-architecture. The former is concerned with the internal structure of experts, whereas the latter is focused on the characteristics of the overall population –including training, input space partitioning, output combination, and evolutionary behavior. Furthermore, the solution adopted to deal with the problem of how to encode inputs for embedded experts is briefly outlined in a separate subsection.

3.1   NXCS Micro-architecture

In its basic form, the general structure of an NXCS expert Γ is a triple $\langle g, h, w \rangle$, where the guard g is a binary function devised to accept or discard inputs according to the value of some relevant features, h is an embedded expert whose activation depends on g, and w is a weighting function, used to perform output combination. Hence, $\Gamma(x)$ coincides with $h(x)$ for any input x that matches $g(x)$, otherwise it is not defined. An expert Γ contributes to the final prediction according to the value of its weighting function w. Conceptually, $w(x)$ represents the expert strength in the voting mechanism and may depend or not on the input x, on the overall fitness of the corresponding expert, and on the reliability of the prediction.

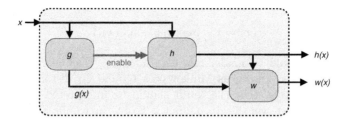

Fig. 1. The micro-architecture of a NXCS expert

The current implementation of NXCS experts is highly configurable and permits several variations on the structure of guards and embedded experts. For the sake of brevity, let us concentrate on the solutions adopted in the task of predicting protein secondary structures (see Figure 1).

Structure of Guards. In the simplest case, the main responsibility of g is to split the input space into matching / non-matching regions (with hard boundaries), with the goal of facilitating the training of h. In a typical evolutionary setting, each guard performs a "hard-matching" activity, implemented by resorting to an embedded pattern in $\{0, 1, \#\}^L$, where "$\#$" denotes the usual "dont-care" symbol and L denotes the length of the pattern. Given an input x, consisting of a string in the alphabet $\{0, 1\}$, the matching between x and g returns *true* if and only if all non-$\#$ values coincide (otherwise, the matching returns *false*). It is trivial to extend this definition by devising guards that map inputs to $[0, 1]$. Though very simple from a conceptual perspective, this relaxed interpretation requires the adoption of a flexible matching mechanism, which has been devised according to the following semantics: Given an input x, a guard g evaluates the overall matching score $g(x)$, and activates the corresponding embedded expert h if and only if $g(x) \geq \theta$ (the threshold θ is a system parameter).

Let us assume that g embeds a pattern e, represented by a string in $\{0, 1, \#\}$ of length L, used to evaluate the distance between an input x and the guard. To improve the generality of the system, one may assume that a vector of relevant, domain dependent, features is provided, able to implement a functional transformation from x to $[0, 1]^L$. In so doing, the i-th feature, denoted by $m_i(x)$, can be associated with the i-th value, say e_i, of the embedded pattern e. Under these assumptions, the function $g(x)$ can be defined as (d denotes a suitable distance metrics):

$$g(x) = 1 - d(e, m(x)) \tag{1}$$

In our opinion the most natural choice for implementing the distance metrics should extend the hard-matching mechanism used in a typical evolutionary setting. In practice, the i-th component of e controls the evaluation of the corresponding input features, so that only non-"$\#$" features are actually taken into account. Hence, $H_g \neq \emptyset$ being the set of all non-"$\#$" indexes in e, $g(x)$ can be defined, according to the Minkowski's L_∞ distance metrics, as:

$$g(x) = 1 - \max_{i \in H_g} \{|e_i - m_i(x)|\} \tag{2}$$

Other choices could be made, too, e.g., adopting Euclidean or Manhattan distance metrics (also known as Minkowski's L_2 and L_1 distance metrics). In either case, let us stress that the result should be interpreted as a "degree of expertise" of an expert over the given input x. To give an example of the matching activity, let us assume that the embedded pattern e of a guard g is defined as:

$$e = \langle \#, 1, \#, \#, 0, \#, ..., \# \rangle$$

In this case, only the second and the fifth feature are active; hence, in this case:

$$g(x) = 1 - \max_{i \in \{2,5\}} \{|e_i - m_i(x)|\} = 1 - \max\{1 - m_2(x), m_5(x)\}$$

It is worth noting that a pattern composed only by dont-care symbols would actually yield an expert with a complete visibility on the input space (i.e., a globally-scoped expert). In this trivial case, we assume that $g(x) = 1$ for each input x. In the following, we make the hypothesis that a typical expert has at least one non-"#" symbol in e (i.e., that the corresponding expert is, at least in principle, locally-scoped).

Structure of Embedded Experts. As for embedded experts, a simple Multi Layer Perceptron (MLP) architecture has been adopted –equipped with a single hidden layer. The issue of the dependence between the number of inputs and the number of neurons in the hidden layer has also been taken into account. In particular, several experimental results addressed the issue of finding a good tradeoff between the need of limiting the number of hidden neurons and the need of augmenting it (to prevent overfitting and underfitting, respectively). Let us stress in advance that the problem of reducing overfitting has been dealt with by adopting a suitable input encoding, which greatly limits the phenomenon. As a consequence, the underfitting problem has also become more tractable, due to the fact that the range of "reasonable" choices for ANN architectures has increased. In particular, an embedded expert with a complete visibility of the input space is equipped with 35 hidden neurons, whereas experts enabled by 10%, 20% and 30% of the input space are equipped with 10, 15, and 20 neurons, respectively.

3.2 NXCS Macro-architecture

A population of multiple experts exhibits a behavior that is completely specified only when suitable implementation choices have been made –typically dependent on the given application task. In our opinion, the most relevant features to take into account for designing systems based on multiple experts are: (a) training, (b) selection mechanisms, and (c) output combination (see also [63]).

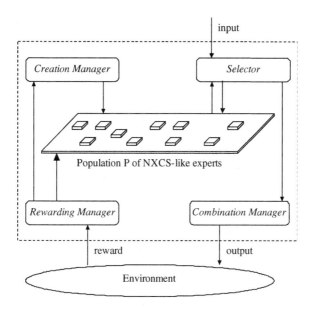

Fig. 2. A population of NXCS experts

Training NXCS Experts. Training occurs in two steps, which consist of (1) discovering a population of guards aimed at soft partitioning the input space, and (2) training the embedded experts of the resulting population.

In the first step, the training technique adopted for NXCS guards is basically conformant to the one adopted for XCS classifier systems, i.e., an accuracy-based selection is enforced on the population of experts according to the reward obtained by the underlying environment. In particular, experts are generated concentrating only on the "partitioning" capability of their guards (let us recall that a guard is aimed at identifying a context able to facilitate the prediction performed by the corresponding embedded expert). In particular, the system starts with an initial population of experts equipped with randomly-generated guards. In this phase, embedded experts play a secondary role, their training being deferred to the second step. Until then, they output only the statistics (in terms of percent of α, β, and c) computed on the subset of inputs acknowledged by their guard, no matter which input x is being processed. Prediction, prediction error, accuracy and fitness are evaluated according to the basic strategy of XCS systems, the reward given for alpha helices, beta sheets, and coils being inversely proportional to their (percent of) occurrence in the training set. Further experts are created according to a covering, crossover, or mutation mechanism. As for covering, given an input x of class c, a guard that matches x is generated by a greedy algorithm driven by the goal of maximizing the percent of inputs that share the class c within the inputs covered by the guard itself. As usual for genetic-based populations, mutation and crossover operate on the embedded patterns of the parent(s), built upon the alphabet $\{0,1,\#\}$. It is worth pointing out that, at the end of the first step, a globally-scoped expert (i.e., equipped with a guard whose embedded pattern contains only "#") is inserted in the population, to guarantee that the input space is completely covered in any case. From this point on, no further creation of experts is performed.

In the second step the focus moves to embedded experts, which, turned into MLPs, are trained using the backpropagation algorithm on the subset of inputs acknowledged by their corresponding guard. In the current implementation of the system, all MLPs are trained in parallel, until a convergence criterion is satisfied or the maximum number of epochs has been reached. The training of MLPs follows a special technique, explicitly devised for this specific application. In particular, given an expert consisting of a guard g and its embedded expert h, h is trained on the whole training set in the first five epochs, whereas the visibility of the training set is restricted to the inputs matched by the corresponding guard in the subsequent epochs. In this way, a mixed training strategy has been adopted, whose rationale lies in the fact that experts must find a suitable trade-off between the need of enforcing diversity (by specializing themselves on a relevant subset of the input space) and the need of preventing overfitting.

Selection Mechanisms. By hypothesis, in any implementation of NXCS systems, experts do not have complete visibility of the input space (i.e., they operate on different regions), which is also a common feature for all approaches based on genetic algorithms. In the implementation designed for predicting protein secondary structures, regions exhibit a particular kind of "soft" boundaries, in accordance with the flexible matching activity performed on any input x. This mechanism makes it possible for experts to show different "ranges of authority" on the input x, which is also a common feature for systems that adopt soft boundaries. As already pointed out, in the current implementation of the system, an expert is selected when the matching activity returns a value greater than a given threshold θ. In this case, a selection mechanism is required, aimed at identifying and collecting all experts that are able to deal with the given input. Given an input x, all selected experts form the match set, denoted by $M(x)$.

Output Combination. Let us recall that each expert outputs three signals in [0,1], representing its "confidence" in predicting the class labels α, β, and c. Once the match set has been formed, output combination occurs by enforcing weighted averaging. In particular, experts support each separate class label according to the value of their corresponding output, modulated by w. In the current implementation of the system, w depends (i) on the degree of matching $g(x)$, (ii) on the expert's fitness, and (iii) on the reliability of the prediction, estimated by the difference between the two highest outputs of $h(x)$ [55]. Given an input x, for each expert $\Gamma \in M(x)$, let us denote with $g_\Gamma(x)$ the value resulting from the matching, with $h_\Gamma^k(x)$ the k-th output ($k \in \{\alpha, \beta, c\}$) of its embedded predictor h, and with $w_\Gamma(x)$ the value of its weighting function w. Under these hypotheses, the system labels the input x according to the decision rule:

$$k^* = \arg\max_{k \in \{\alpha, \beta, c\}} \left\{ O^k(x) \right\}$$

where:

$$O^k(x) = \frac{\sum\limits_{\Gamma \in M(x)} h_\Gamma^k(x) \cdot w_\Gamma(x)}{\sum\limits_{\Gamma \in M(x)} w_\Gamma(x)} \qquad k \in \{\alpha, \beta, c\}$$

and, f_Γ and $r_\Gamma(x)$ being the fitness and the reliability of the expert Γ:

$$w_\Gamma(x) = f_\Gamma \cdot g_\Gamma(x) \cdot r_\Gamma(x)$$

Implementing the NXCS Evolutionary Behavior. To facilitate the implementation of the evolutionary behavior, auxiliary software agents have been devised and implemented: (i) a creation manager, (ii) a selector, (iii) a combination manager, and (iv) a rewarding manager (see Figure 2). The *creation manager* is responsible for creating experts. Whilst during the initialization phase the creation manager follows a random strategy, afterwards it is invoked to decide which experts must undergo crossover or mutation. The creation manager is also invoked when the current input is not covered by any existing expert. In this case, an expert able to cover the given input is created on-the-fly. The *selector* is devoted to collect all experts whose guard matches the given input x, thus forming the match set $M(x)$. The combination manager is entrusted with combining the outputs of experts belonging $M(x)$, i.e., it applies the voting policy described above. The main task of the *rewarding manager* is forcing all experts in $M(x)$ to update their internal parameters, in particular the fitness, according to the reward obtained by the external environment. The rewarding manager is also responsible for deciding which experts should be deleted (only during the first step of the training activity); in particular, experts whose fitness is under a given threshold will be deleted, if needed, with a probability inversely proportional to their fitness.

3.3   Input Encoding for Embedded Experts

As for the problem of identifying a suitable encoding aimed at facilitating the prediction, in our opinion, most of the previous work, deeply rooted in the ANN technology, can be revisited to better highlight the link existing between the ability of dealing with inputs encoded using a one-shot technique and the adoption of multialignment. In fact, we claim that the impressive improvements in prediction performance accounted for the adoption of multialignment (see, for example, [55]) are not only due to the injection of further relevant information, but are also strictly related with the ANN technology and its controversial ability of dealing with one-shot encodings. This issue has been carefully analyzed, and experiments have been performed with the subgoal of highlighting the relationship between one-shot encoding and multialignment. Experimental results made with one-shot encoding point out that it prevents the ANNs from performing generalization on the given task. In the past, several attempts have been made to overcome this problem. In particular, Riis and Krogh [52] have shown that the percent of correct prediction can be greatly improved by learning a three-dimensional encoding for aminoacids, without resorting to multialignment. Their result is an indirect confirmation of our conjecture, which assumes that multialignment techniques for encoding inputs can improve the performance of a system, not only due to the injection of additional information, but also thanks to their capability of dealing with the one-shot problem. Going a step further in the latter direction, we propose a solution based on the *Blosum80* [29] substitution matrix. In fact, being averaged on a great number of proteins, we deem that the information contained in the *Blosum80* matrix contains a kind of "global" information that can be used to contrast the drawbacks of the one-shot encoding. In order to highlight the proposed solution, let us give some preliminary definition:

- Each aminoacid is represented by an index in [1-20] (i.e., 1/Alanine , 2/Arginine, 3/Asparagine, ..., 19/Tyrosine, 20/Valine). The index 0 is reserved for representing the gap.
- $\mathbf{P} = \langle P_i, i = 0, 1, ..., n \rangle$ is a list of sequences where (i) P_0 is the protein to be predicted (i.e. the primary input sequence), containing L aminoacids, and (ii) $P_i, i = 1, ..., n$ is the list of sequences related with P_0 by means of similarity-based metrics, retrieved using BLAST. Being multialigned with P_0, these sequences usually contain gaps, so that their length still amounts to L. Furthermore, let us denote with $P(j), j = 1, 2, ..., L$ the j-th column of the multialignment, and with $P_i(j), j = 1, 2, ..., L$ the j-th residue of the sequence P_i.
- $\mathbf{B}$ is a 21 × 21 matrix obtained by normalizing the *Blosum80* matrix in the range [0,1]. Thus, B_k denotes the row of $\mathbf{B}$ that encodes the aminoacid k ($k = 1, 2, ..., 20$), whereas $B_k(r)$ represents the degree of substitability of the r-th aminoacid with the k-th aminoacid. The row and the column identified by the 0-th index represent the gap, set to a null vector in both cases –except for the element $B_0(0)$ which is set to 1.
- $\mathbf{Q}$ is a matrix of 21 × L positions, representing the final encoding of the primary input sequence P_0. Thus, $Q(j)$ denotes the j-th column of the matrix, which is intended to encode the j-th amino acid (i.e., $P_0(j)$) of the primary input sequence (i.e., P_0), whereas $Q_r(j), r = 0, 1, ..., 20$ represents the contribution of the r-th aminoacid in the encoding of $P_0(j)$ (the index $r = 0$ is reserved for the gap).

The normalization of the *Blosum80* matrix in the range [0,1], yielding the $\mathbf{B}$ matrix, is performed according to the following guidelines:

1. μ and σ being the mean and the standard deviation of the *Blosum80* matrix, respectively, calculate the "equalized matrix" $\mathbf{E}$ by applying a suitable sigmoid function, whose zero-crossing is set to μ and with a range in $[-\sigma, \sigma]$. In symbols:

$$\forall k = 1, 2, ..., 20 : \forall j = 1, 2, ..., 20 : E_k(j) \leftarrow \sigma \cdot \tanh(Blosum80_k(j) - \mu)$$

2. E^m and E^M being the minimum and the maximum value of the equalized matrix $\mathbf{E}$, respectively, build the normalized matrix $\mathbf{B}$. In symbols (the 0-th row and column of $\mathbf{B}$ are used to encode gaps):

$$B_0 \leftarrow \langle 1, 0, ..., 0 \rangle, \ B(0) \leftarrow \langle 1, 0, ..., 0 \rangle^T$$

$$\forall k = 1, 2, ..., 20 : \forall j = 1, 2, ..., 20 : B_k(j) \leftarrow \frac{E_k(j) - E^m}{E^M - E^m}$$

It is worth pointing out that the evaluation of $\mathbf{B}$ occurs once, being independent from any particular sequence to be processed.

Starting from the above definitions, the algorithm used for encoding the primary input sequence P_0 is:

1. Initialize $\mathbf{Q}$ with the *Blosum80*-like encoding of the primary sequence P_0 (B_s^T represents the vector B_s transposed). In symbols:

$$\forall j = 1, 2, ..., L : s \leftarrow P_0(j), \ Q(j) \leftarrow B_s^T$$

2. Update $\mathbf{Q}$ according to the *Blosum80*-like encoding of the remaining sequences $P_1, P_2, ..., P_n$. In symbols:

$$\forall i = 1, 2, ..., n : \forall j = 1, 2, ..., L : s \leftarrow P_i(j), \ Q(j) \leftarrow Q(j) + B_s^T$$

3. Normalize the elements of $\mathbf{Q}$, column by column, in [0,1]. In symbols:

$$\forall j = 1, 2, ..., L : \gamma \leftarrow \sum_s Q_s(j), \ \forall r = 0, 1, 2, ..., 20 : Q_r(j) \leftarrow Q_r(j)/\gamma$$

According to our experimental results, the encoding defined above greatly contributes to reduce overfitting and produces an improvement of about 1.5% in the prediction performance. Although not deeply rooted in statistical theory, the adopted input encoding makes a step further in the direction of removing the drawback introduced by the classical one-shot encoding, also due to the fact that the *Blosum80* matrix is logarithmic. Furthermore, it is worth noting that also in this case a mixed strategy –in a sense similar to the one adopted for training ANNs– has been enforced, where the information contained in the *Blosum80* matrix and in multiple alignment represent the "global" and the "local" part, respectively.

4   Experimental Results

To assess the performance of the predictor, also facilitating a comparison with other systems, we adopted the TRAIN and the R126 datasets, for training and testing, as described in [51]. The TRAIN dataset has been derived from a PDB selection obtained by removing short proteins (less than 30 aminoacids), and with a resolution of at least 2.5 Å. This dataset underwent an homology reduction, aimed at excluding sequences with more than 50% of similarity. Furthermore, proteins in this set have less than 25% identity with the sequences in the set R126. The resulting training set consists of 1180 sequences, corresponding to 282,303 amino acids. The R126 test dataset is derived from the historical Rost and Sander's protein dataset (RS126) [55], and corresponds to a total of 23,363 amino acids (the overall number has slightly varied over the years, due to changes and corrections in the PDB).

Guards have already been described in great detail in a separate section. The list of features handled by guards (adopted for soft partitioning the input space) is reported in Table 1. As already pointed out, embedded predictors are in fact MLPs, equipped with one hidden layer containing a number of neurons (i.e., 10, 15, 20 or 35) that depends on the size of the input space that –at least in principle– can be processed by the expert. The input window r, to be moved along proteins, consists of 15 contiguous residues (i.e., 7+1+7 residues in

Table 1. Features used for partitioning the input space (each feature is evaluated on a window of length r and centered around the residue to be predicted)

	Feature	Conjecture
1	Check whether hydrophobic aminoacids occur in the current window ($r=15$) according to a clear periodicity (i.e., one every 3-4 residues)	Alpha helices may sometimes fulfil this pattern
2	Check whether the current window ($r=13$) contains numerous residues in {A,E,L,M} and few residues in {P,G,Y,S}	Alpha helices are often evidenced by {A,E,L,M} residues, whereas {P,G,Y,S} residues account for their absence
3	Check whether the left side of the current window ($r=13$) is mostly hydrophobic and the right part is mostly hydrophilic (and viceversa)	Transmembrane alpha helices may fulfil this feature
4	Check whether, on the average, the current window ($r=11$) is positively charged or not	A positive charge might account for alpha helices or beta sheets
5	Check whether, on the average, the current window ($r=11$) is negatively charged or not	A negative charge might account for alpha helices or beta sheets
6	Check whether, on the average, the current window ($r=11$) is neutral	A neutral charge might account for coils
7	Check whether the current window ($r=11$) mostly contains "small" residues	Small residues might account for alpha helices or beta sheets
8	Check whether the current window ($r=11$) mostly contains polar residues	Polar residues might account for alpha helices or beta sheets

the window). Each residue is encoded using 21 real values in $[0, 1]$ calculated in accordance with the encoding algorithm described in the previous section.

In the experiments carried out, the population was composed by 600 experts, with about 20 experts (on average) involved in the match set. The threshold θ has been set to 0.4. As for MLPs, the learning rate has been set to 0.07 and the number of epochs to 80.

In order to evaluate how performance depends on the amount of domain knowledge injected into the system, we compared the performance obtained by adopting random guards (Figure 3) with those obtained by means of a genetic selection (Figure 4).

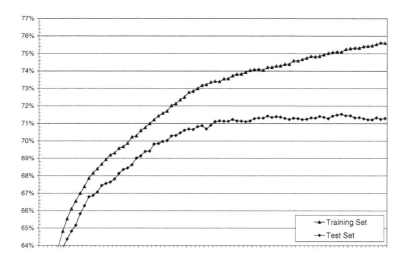

Fig. 3. The overall performance of MASSP, run with randomly-generated guards, obtained while training embedded MLP predictors, are reported at different epochs. The maximum value in the test set is 71.6%

The results obtained in the above experimental settings point out that the evolutionary setting performs better (74.8%) than the one characterized by randomly-generated guards (71.6%). In our opinion, the difference of about 3% is due to the domain knowledge embodied by the most "successful" experts, despite the fact that the quality of the adopted features has not been assessed by a biologist. Our conjecture is that more "biologically-biased" features would cause a further improvement in prediction performance.

To facilitate the comparison with other relevant systems, MASSP has also been assessed according to the guidelines described in [20]. In particular, the programs NNSSP, PHD, DSC, PREDATOR, CONSENSUS have been considered concerning performance against the commonly used RS126 dataset. Having trained MASSP using the same data set (i.e., TRAIN), we reported also the performance of SSPRO [51]. Experimental results are summarized in Table 2.

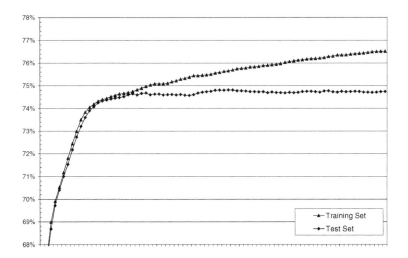

Fig. 4. The overall performance of MASSP, run with genetically selected guards, obtained while training embedded MLP predictors, are reported at different epochs. The maximum value in the test set is 74.8%.

Table 2. Experimental results, obtained from the RS126 dataset

System	Q3
PREDATOR	70.3
DSC	71.1
NNSSP	72.7
PHD	73.5
CONSENSUS	74.8
MASSP	**74.8**
SSPRO	76.6

As a final remark on experimental results, let us point out that the focus of the paper was mainly on the impact of the evolutionary setting (that permits to evolve useful combinations of domain-oriented features) on the overall performance of MASSP. The fact that SSPRO obtains better results is not surprising, this system being based on a techology (i.e., Recurrent ANNs –see, for instance, [14]) which is more adequate than MLPs for processing sequences.

5   Conclusions and Future Work

In this paper, an approach for predicting protein secondary structures has been presented, which relies on a multiple-expert architecture. In particular, a population of hybrid experts –embodying a genetic and a neural part– has been suitably devised to perform the given application task. Experimental results, performed on sequences taken from well-known protein databases, point to the validity of the approach. As for the future work, we are about to test a post-processor devoted to improve the overall performance of the system by means of a secondary-to-secondary structure prediction. Furthermore, in collaboration with a biologist, we are trying to devise more "biologically-based" features – to be embedded in genetic guards– able to improve their ability of performing context identification. The adoption of RANNs is also being investigated as the underlying technology for implementing embedded experts.

References

1. Altschul, S.F., Gish, W., Miller, W., Myers, E.W., Lipman, D.J.: Basic local alignment search tool. J. Mol. Biol. (1990) 215:403–10
2. Altschul, S.F., Madden, T.L., Schaeffer, A.A., Zhang, J., Zhang, Z., Miller, W., Lipman, D.J.: Gapped BLAST and PSI-BLAST: a new generation of protein database search programs. Nucleic Acids Res. (1997) 25:3389–3402
3. Anfinsen, C.B.: Principles that govern the folding of protein chains. Science. (1973) 181:223–230
4. Armano, G.: NXCS Experts for Financial Time Series Forecasting. in Applications of Learning Classifier Systems, Larry Bull (ed.), Springer (2004) 68–91
5. Bairoch A., Apweiler R.: The SWISS-PROT protein sequence database and its supplement TrEMBL in 2000. Nucleic Acids Res. (2000) 28:45–48
6. Baldi, P., Brunak, S., Frasconi, P., Soda, G., Pollastri, G.: Exploiting the Past and the Future in Protein Secondary Structure Prediction. Bioinformatics. (1999) 15:937–946
7. Baldi, P., Brunak, S., Frasconi, P., Pollastri, G., Soda, G.: Bidirectional Dynamics for Protein Secondary Structure Prediction. In Sun, R., Giles, C.L. (eds.): Sequence Learning: Paradigms, Algorithms, and Applications. Springer-Verlag, (2000) 80-104
8. Berman, H.M., Westbrook, J., Feng, Z., Gilliland, G., Bhat, N., Weissig, H., Shindyalov, I.N., Bourne, P.E.: The Protein Data Bank. Nucleic Acids Research. (2000) 28:235–242
9. Blundell, T.L., Johnson, M.S.: Catching a common fold. Prot. Sci. (1993) 2(6):877–883
10. Boczko, E.M., Brooks, C.L.: First-principles calculation of the folding free energy of a three-helix bundle protein. Science. (1995) 269(5222):393–396
11. Bowie, J.U., Luthy, R., Eisenberg, D.: A method to identify protein sequences that fold into a known 3-dimensional structure. Science (1991) 253:164–170
12. Breiman, L., Friedman, J., Olshen, R., Stone, C.: Classification and Regression Trees. Wadsworth, Belmont CA (1984)
13. Breiman, L.: Stacked Regressions. Machine Learning (1996) 24:41–48
14. Cleeremans, A.: Mechanisms of Implicit Learning. Connectionist Models of Sequence Processing. MIT Press (1993)
15. Chothia, C., Lesk, A.M.: The relation between the divergence of sequence and structure in proteins. EMBO J., (1986) 5:823–826
16. Chothia, C.: One thousand families for the molecular biologist. Nature (1992) 357:543–544
17. Chou, P.Y., Fasman, U.D.: Prediction of protein conformation. Biochem. (1974) 13:211–215

18. Chothia, C.: Proteins – 1000 families for the molecular biologist. Nature (1992) 357:543–544

19. Clark, P., Niblett, T.: The CN2 Induction Algorithm. Machine Learning (1989) 3(4):261–283

20. Cuff, J.A., Barton,G.J.: Evaluation and improvement of multiple sequence methods for protein secondary structure prediction. PROTEINS: Structure, Function and Genetics (1999) 34:508–519.

21. Dandekar, T., Argos., P.: Folding the main chain of small proteins with the genetic algorithm. J. Mol. Biol., (1994) 236:844–861

22. Covell, D.G.: Folding protein alpha-carbon chains into compact forms by Monte Carlo methods. Proteins (1992) 14:409–420

23. Flockner, H., Braxenthaler, M., Lackner, P., Jaritz, M., Ortner, M. Sippl, M.J.: Progress in fold recognition. Proteins: Struct., Funct., Genet. (1995) 23:376–386

24. Freund, Y., Schapire, R.E.: A Decision-Theoretic Generalization of On-Line Learning and an Application to Boosting. Journal of Computer Science and System Sciences (1997) 55(1):119–139

25. Gething, M.J., Sambrook, J.: Protein folding in the cell Nature (1992) 355:33–45

26. Goldberg, D.E.: Genetic Algorithms in Search, Optimization and Machine Learning. Addison-Wesley (1989)

27. Greer, J.: Comparative modelling methods: application to the family of the mammalian serine proteases. Proteins (1990) 7:317–334

28. Havel, T.F.: Predicting the structure of the flavodoxin from Eschericia coli by homology modeling, distance geometry and molecular dynamics. Mol. Simulation, (1993) 10:175–210

29. Henikoff, S., Henikoff, J. G.: Amino acid substitution matrices from protein blocks. Proc. Nat. Acad. Sci. (1989), 10915–10919.

30. Holley, H.L., Karplus, M.: Protein secondary structure prediction with a neural network. Proc. Natl. Acad. Sc., U.S.A. (1989) 86:152–156

31. Hartl, F.U.: Secrets of a double-doughnut. Nature (1994) 371:557–559

32. Higgins, D., Thompson, J., Gibson T., Thompson, J.D., Higgins, D.G., Gibson, T.J.: CLUSTAL W: improving the sensitivity of progressive multiple sequence alignment through sequence weighting, position-specific gap penalties and weight matrix choice. Nucleic Acids Res. (1994) 22:4673–4680

33. Hinds, D.A., Levitt, M.: Exploring conformational space with a simple lattice model for protein structure. J. Mol. Biol. (1994) 243:668–682

34. Holland, J.H.: Adaptation in Natural and Artificial Systems. University of Michigan Press (1975)

35. Holland, J.H.: Adaption. In: Rosen, R., Snell, F.M. (eds.): New York Academic Press. Progress in Theoretical Biology (1976) 4:263–293

36. Holland, J.H.: Escaping Brittleness: The possibilities of General-Purpose Learning Algorithms Applied to Parallel Rule-Based Systems. In: Michalski, R.S., Carbonell, J., Mitchell, M. (eds.): Machine Learning, An Artificial Intelligence Approach, (1986) vol II 20:593–623 Morgan Kaufmann

37. Jacobs, R.A., Jordan, M.I., Nowlan, S.J., Hinton, G.E.: Adaptive Mixtures of Local Experts. Neural Computation (1991) 3:79–87

38. Jones, D. T., Taylor, W. R. and Thornton, J. M.: A new approach to protein fold recognition. Nature (1992) 358:86-89.

39. Jones, D.T.: Protein secondary structure prediction based on position-specific scoring matrices. J. Mol. Biol. (1999) 292:195–202

40. Jordan, M.I., Jacobs, R.A.: Hierarchies of Adaptive Experts. In Moody, J., Hanson, S., Lippman, R. (eds.): Advances in Neural Information Processing Systems (1992) 4:985-993 Morgan Kaufmann

41. Kanehisa, M.: A multivariate analysis method for discriminating protein secondary structural segments. Prot. Engin. (1988) 2:87–92

42. Krogh, A., Vedelsby, J.: Neural Network Ensembles, Cross Validation, and Active Learning. In Tesauro, G., Touretzky, D., Leen, T. (eds.): MIT Press Advances in Neural Information Processing Systems (1995) vol 7 231–238

43. Lathrop, R.H. Smith, T.F.: Global optimum protein threading with gapped alignment and empirical pair score functions. J. Mol. Biol. (1996) 255:641–665

44. Levitt, M.: Protein folding by constrained energy minimization and molecular dynamics. J. Mol. Biol. (1983) 170:723–764

45. Levitt, M.: A simplified representation of protein conformations for rapid simulation of protein folding. J. Mol. Biol. (1976) 104:59–107

46. Madej, T., Gibrat, J.F., Bryant, S.H.: Threading a database of protein cores. Proteins: Struct., Funct., Genet. (1995) 23:356–369

47. Mitchell, E.M., Artymiuk, P.J., Rice, D.W., Willett, P.: Use of techniques derived from graph theory to compare secondary structure motifs in proteins. J. Mol. Biol. (1992) 212:151–166

48. Orengo, C.A., Jones, D.T., Thornton, J.M.: Protein superfamilies and domain superfolds. Nature (1994) 372:631–634

49. Quinlan, J.R.: Induction of Decision Trees. Machine Learning (1986) 1:81–106

50. Ptitsyn, O.B., Finkelstein, A.V.: Theory of protein secondary structure and algorithm of its prediction. Biopolymers (1983) 22:15–25

51. Pollastri, G., Przybylski, D., Rost, B., Baldi, P.: Improving the Prediction of Protein Secondary Structure in Three and Eight Classes Using Neural Networks and Profiles. Proteins (2002) 47:228–235

52. Riis, S.K., Krogh, A.: Improving prediction of protein secondary structure using structured neural networks and multiple sequence alignments. J. Comp. Biol. 3 (1996), 163–183.

53. Rivest, R.L.: Learning Decision Lists. Machine Learning (1987) 2(3):229–246

54. Robson, B.: Conformational properties of amino acid residues in globular proteins. J. Mol. Biol. (1976) 107:327–356

55. Rost, B., Sander, C.: Prediction of protein secondary structure at better than 70% accuracy. J Mol Biol (1993) 232:584-599

56. Roterman, I.K., Lambert, M.H., Gibson, K.D., Scheraga, H.A.: A comparison of the charmm, amber and ecepp potentials for peptides. ii. phi-psi maps for n-acetyl alanine n'-methyl amide: comparisons, contrasts and simple experimental tests. J. Biomol. Struct. Dynamics, (1989) 7:421–453

57. Russell, R.B., Copley, R.R. Barton, G.J.: Protein fold recognition by mapping predicted secondary structures. J. Mol. Biol. (1996) 259:349–365

58. Sali, A.: Modelling mutations and homologous proteins. Curr. Opin. Biotech. (1995) 6:437-451

59. Salamov, A.A., Solovyev, V.V.: Prediction of protein secondary structure by combining nearest-neighbor algorithms and multiple sequence alignment. J. Mol. Biol. (1995) 247:11–15

60. Sanchez R., Sali, A.: Advances in comparative protein-structure modeling. Curr. Opin. Struct. Biol. (1997) 7:206–214

61. Schapire, E.: A Brief Introduction to Boosting. Proc. of the Sixteenth Int. Joint Conference on Artificial Intelligence (1999) 1401–1406

62. Skolnick, J., Kolinski, A.: Simulations of the folding of a globular protein. Science, (1990) 250:1121–1125

63. Sun, R., Peterson, T.: Multi-agent reinforcement learning: weighting and partitioning. Neural Networks (1999) vol 12 4-5:127–153

64. Taylor, W.R., Thornton, J.M.: Prediction of super-secondary structure in proteins. Nature (1983) 301:540–542

65. Taylor, W.R., Orengo, C.A.: Protein-structure alignment. J. Mol. Biol. (1989) 208:1–22.

66. Unger, R., Harel, D., Wherland, S., Sussman, J.L.: A 3-D building blocks approach to analyzing and predicting structure of proteins. Proteins (1989) 5:355–373

67. Vajda, S., Sippl, M., Novotny, J.: Empirical potentials and functions for protein folding and binding. Curr. Opin. Struct. Biol. (1997) 7:228–228

68. Valiant L.: A Theory of the Learnable. Communications of the ACM (1984) 27:1134–1142

69. Vapnik, V.N.: Statistical Learning Theory. John Wiley and Sons Inc., New York (1998)

70. Vere, S.A.: Multilevel Counterfactuals for Generalizations of Relational Concepts and Productions. Artificial Intelligence (1980) 14(2):139–164

71. Weigend, A.S., Mangeas, M., Srivastava, A.N.: Nonlinear Gated Experts for Time Series: Discovering Regimes and Avoiding Overfitting. Int. Journal of Neural Systems, 6(1995):373-399

72. Wilson, S.W.: Classifier Fitness Based on Accuracy. Evolutionary Computation (1995) 3(2):149–175

Modeling Kohn Interaction Maps with Beta-Binders: An Example[*]

Federica Ciocchetta, Corrado Priami, and Paola Quaglia

Dipartimento di Informatica e Telecomunicazioni,
Università di Trento, Italy

Abstract. We represent a subset of the mammalian cell cycle Kohn interaction map using Beta-binders, a formalism inspired to the pi-calculus and enriched with binders that allow the description of enclosing surfaces equipped with interaction sites. Specifically, the interactions between the p53 protein and its main regulator Mdm2 is analyzed.

Beta-binders comes equipped with a reduction semantics for the formal description of the evolution of the specified systems. This allows a *dynamic* representation of the intrinsically *static* Kohn maps.

1 Introduction

The rapid progress in molecular biology has produced a large quantity of biological data and has led to the understanding of numerous interactions and relationships among a lot of biological entities. In particular, it has been possible to acquire a deep knowledge in complex protein and gene networks. A main challenge now is to organize and represent the known interactions in an exhaustive way. Yet another and possibly bigger challenge is to provide foundational models to mechanized tools which can aid "in silico" predictive research on evolutionary behaviour, or just on the behaviour of biological systems whose components are not throughly investigated and understood.

As a response to the need of modeling the dynamics of biological systems, a number of *ad hoc* process calculi have been recently proposed by the research community in concurrency theory (see, e.g., [13,15,3,14,2,4,11]). Generically speaking, they provide the means to specify and reason about protein interactions and complex protein pathways. Nonetheless, each of them has been conceived to respond to some specific modelization concern, and it is yet not clear which of them – if any – can be considered as a general model to formally reason about biology.

To acquire confidence in the expressive power of any formal language it is necessary to test it against realistic case studies. In this paper we push towards this direction by applying one of the above mentioned formalisms to specify and describe (part of) the dynamic behaviour of the p53 protein, as it was graphically illustrated by Kohn in [6]. The formalism we use, called Beta-binders [11], is

---

[*] This work has been partially supported by the FIRB project "Modelli formali per Sistemi Biochimici".

C. Priami et al. (Eds.): Trans. on Comput. Syst. Biol. III, LNBI 3737, pp. 33–48, 2005.
© Springer-Verlag Berlin Heidelberg 2005

strongly inspired by the π-calculus [9,16], and hence takes communication as a primitive form of interaction between entities running (i.e. living) in parallel. The basic ingredient of communication is synchronization of input and output actions on named channels (as in π-calculus 'names' are synonyms of 'channels'). Moreover, Beta-binders provides the means to model the enclosing surfaces of entities and possible interactions taking place at the level of the virtual surfaces. This is due to a special class of binders which are used to wrap (quasi-) π-calculus processes into *boxes* with interaction capabilities. Specifically, boxes represent the borders of biological entities and are equipped with typed interaction sites. The graphical representation of a simple process is shown below.

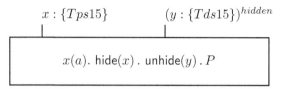

The above box abstractly represents a protein. It can interact with the outside world (a collection of other boxes) through the site named x. In particular, by executing the input action $x(a)$, the protein can undergo phosphorylation. After that, the phosphorylation site is hidden (action $\mathsf{hide}(x)$), and a site named y is made visible on the surface of the box (action $\mathsf{unhide}(y)$). This latest site denotes readiness to dephosphorylation, which may take place or not, depending on the semantics associated with process P and on the specification of the surrounding boxes.

More generally, the evolution of boxes is described by a limited number of macro-operations: communication between components within the same box or within distinct boxes (intra-communication or inter-communication, respectively); addition of a site to the interface; hiding and unhiding of an interaction site; joining of two boxes; splitting of a box in two.

In this paper we apply Beta-binders to model a simple Kohn interaction map. A Molecular Interaction Map (MIM) is a static diagram that describes the interactions among molecular species and gives a compact definition and graphical representation of biological networks. Protein and genes are nodes of the diagram and are connected to each other with several sorts of lines, representing distinct kinds of relationships. W.r.t. other maps, MIMs are visually unambiguous and provide symbol conventions for a full range of biological reactions and interactions. The application presented in this paper is taken from the mammalian cell cycle proposed by Kohn in [6]. Specifically, we analyze a sub-map relative to p53, a protein with a significative role in cell cycle control [1,7].

The main gain in translating Kohn maps into Beta-binders processes is that this latest representation allows formal reasoning about the *dynamics* of the specified systems. Also, while MIMs are intrinsically *static* representations of biological phenomena, the formal underpinning of Beta-binders can be the basis for the development of automatic tools to aid in the inference of the dynamic evolution of the described systems. Indeed, a prototype of such kind of tools has already been implemented [5].

The rest of the paper is organized as follows. In the next section we present a short description of Kohn MIMs, the graphical conventions used in interaction maps, and a summary of the role of the p53 protein in the regulation and control of the cell cycle. Section 3 reports an overview on Beta-binders. Then Section 4 presents a translation in Beta-binders of a simple sub-map of the mammalian cell cycle appeared in [6], the one relative to the p53 protein and its interactions with Mdm2. Section 5 concludes the presentation with some final remarks.

2   Kohn Interaction Maps and P53

This section reports a brief description of the biological problem we consider and an introduction to Kohn interaction maps.

Kohn MIMs are a quite successful attempt to describe a molecular interaction system in all its complexity. The main good point of Kohn maps is that they provide compact graphical representations of biological networks which are often vast and extremely complex.

Using MIMs, Kohn reported in [6] an exhaustive description of the mammalian cell cycle and DNA repair. In the present paper, we focus on a specific subset of the above map and model its dynamics through Beta-binders. In particular, we are interested in the representation of the regulation of the p53 protein and DNA repair. The p53 protein plays an important role in the cell cycle control and apoptosis. It is rich in modification sites, domains, and interconnections with other biological entities, and it is also well-known as a tumor suppressor (for this reason, indeed, p53 is sometimes named the "Guardian of the Genome" [8]).

2.1   Protein P53 and Cell Cycle Control

The p53 protein is involved in a lot of interactions with other species and in a multiplicity of biological processes. The three main functions of p53 are, respectively, growth arrest, DNA repair, and apoptosis (cell death). In this presentation we focus on the interaction of p53 with its regulator Mdm2.

In mammalian cells, DNA damage leads to the activation of the gene regulatory protein p53. This protein is normally present in low concentration, and it is negatively regulated by Mdm2. The interaction between p53 and Mdm2 creates a negative feedback loop: p53 is able to activate Mdm2, and Mdm2 can bind to the N-terminus of p53, repress its trans-activation activity and eventually target p53 to proteasome-mediated degradation, acting as a ubiquitin ligase.

When DNA is damaged, some protein kinases, which are able to phoshorylate p53 (e.g., DNA-PK[1] and ATM[2]), are activated. Phosphorylation of p53 may happen at distinct sites of the protein (e.g., at serine15 and serine37). This blocks the possible binding of p53 to Mdm2. As a result of the binding block, p53

---

[1] DNA-dependent casein kinase.
[2] Ataxia Talengiectasia Mutated gene/protein.

accumulates to high levels and stimulates the transcription of several genes, each with its specific role in the cell life cycle. Among them, the gene that encodes the p21 protein has a particularly important function: when complexed with some cyclin-dependent kinases (Cdks), it causes the arrest of the cell growth, so preventing the replication of damaged DNA. Another gene that can be activated by the binding block described above is the Bax gene. Its activation leads to the apoptosis of the cell, and then it stops the proliferation of cells containing abnormal DNA.

After damaged DNA is repaired, kinases are no longer active. The p53 protein is then quickly dephosphorylated and destroyed by Mdm2, and it returns to low-concentration levels.

2.2   Kohn Map Conventions

Kohn maps offer a graphical representation of complex biological networks. Each molecule is drawn as a circle or sometimes as a rectangle (see, for instance, the promoter element in Figure 2 below). Each element occurs only once in the diagram and its potential interactions with other elements are indicated by lines connecting the interested actors. Several symbols are introduced to represent the various interactions in which a given biological entity, as a gene or a protein, may be involved. Examples of easily representable interactions are: multi-protein complexes (both covalent and non-covalent binding), protein modifications (e.g., acetylation and phosphorylation), enzymatic reactions, stimulation and activation of biological processes, and competing interactions.

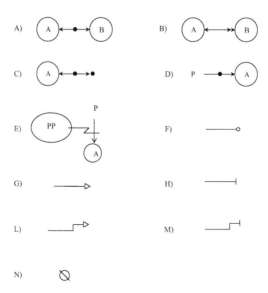

Fig. 1. Kohn map symbols

The complete and detailed list of symbols used in MIMs can be found in [6]. Those symbols which are relevant to the example we analyze are reported in Figure 1 and commented on below.

- *Binding.* The binding between molecular species is indicated by lines terminated at both ends by barbed arrowheads. Barbed lines are symmetric in case of covalent bindings (A). In case of asymmetric bindings the receptor end of the interaction line is represented by a double-barbed arrowhead (B). A bullet in the middle of the connecting line indicates the formed complex. Multiple actions of the complex can be conveniently depicted by using multiple nodes on the same line; each node refers exactly to the same complex. To represent alternative bindings of different proteins at the same site, the line from the competing proteins are merged before the connection to the site.
- *Homopolymer formations.* A complex containing multiple copies of the same monomer is drawn by using a ditto symbol, consisting of an isolated filled circle at the end of a single connecting line (C). The points indicate a copy of the element.
- *Covalent modifications and cleavage of covalent bonds.* To illustrate covalent modifications (e.g., phosphorylation and acetylation) a line with a single barbed arrowhead that points to the modified protein is drawn (D). Dephosphorylation of A by a phosphatase PP is graphically denoted as in (E).
- *Stimulation, inhibition, transcriptional activation and transcriptional inhibition.* Various sorts of lines are used to visualize distinct phenomena. Each line ends with a particular tip to indicate a different kind of interaction. Stimulation by an enzyme is rendered as in (F), generic stimulation as in (G), inhibition as in (H), transcriptional activation as in (L), and transcriptional inhibition as in (M).
- *Degradation.* Degradation products are drawn as slashed circles (N).

3   Beta-Binders

Beta-binders is a formalism strongly inspired by the π-calculus, and equipped with primitives for representing biological compartments and for handling them. This section presents an overview of Beta-binders. For further details and intuitions about the formalism, the interested reader is referred to [11,12].

The π-calculus is a process calculus where names are the media and the values of communication. The same point of view is taken in Beta-binders, where the existence of a countably infinite set N of names is assumed (names are ranged over by lower-case letters). Beta-binders allows the description of the behaviour of π-calculus processes wrapped into boxes with interaction capabilities (hereafter called *beta-processes* or simply *boxes*). Processes within boxes (ranged over by capital letters distinct from B) are given by the following syntax.

$$P ::= \mathsf{nil} \mid x(w).P \mid \overline{x}\langle y\rangle.P \mid P \mid P \mid \nu y\, P \mid !P \mid$$

$$\mathsf{expose}(x,\, \varGamma).P \mid \mathsf{hide}(x).P \mid \mathsf{unhide}(x).P$$

For simplicity, despite the difference w.r.t. the usual π-calculus syntax, we refer to the processes generated by the above grammar as to pi-processes. The deadlocked process nil, input and output prefixes ($x(w).P$ and $\overline{x}\langle y\rangle.P$, respectively), parallel composition ($P \mid P$), restriction ($\nu y\,P$), and replication ($!P$) have exactly the same meaning as in the π-calculus.

The expose, hide, and unhide prefixes are intended for changing the external interface of boxes by adding a new site, hiding a site, and unhiding a site, respectively.

The π-calculus definitions of *name substitution* and of *free* and *bound names* (denoted by fn(-) and bn(-), respectively) are extended to the processes generated by the above syntax in the obvious way. It is sufficient to state that neither hide(x) nor unhide(x) act as binders for x, while the prefix expose(x, Γ) in expose(x, Γ).P is a binder for x in P.

Beta-processes are defined as pi-processes prefixed by specialized binders that suggest the name of the formalism and are defined as follows.

Definition 1. *An* elementary beta-binder *has either the form* $\beta(x : \Gamma)$ *or the form* $\beta^h(x : \Gamma)$, *where*

1. *the name x is the* **subject** *of the beta-binder, and*
2. *Γ is the* **type** *of x. It is a non-empty set of names such that $x \notin \Gamma$.*

Intuitively, the elementary beta-binder $\beta(x : \Gamma)$ is used to denote an active (potentially interacting) site of the box. A binders like $\beta^h(x : \Gamma)$ denotes a site that has been hidden to forbid further interactions through it.

Definition 2. Composite beta-binders *are generated by the following grammar:*

$$\boldsymbol{B} ::= \beta(x : \Gamma) \ \bigg| \ \beta^h(x : \Gamma) \ \bigg| \ \beta(x : \Gamma)\,\boldsymbol{B} \ \bigg| \ \beta^h(x : \Gamma)\,\boldsymbol{B}$$

A composite beta-binder is said to be **well-formed** *when the subjects of its elementary components are all distinct. We let well-formed beta-binders be ranged over by* $\boldsymbol{B}, \boldsymbol{B}_1, \boldsymbol{B}_2, \ldots, \boldsymbol{B}', \ldots$.

The set of the subjects of all the elementary beta-binders in $\boldsymbol{B}$ is denoted by sub($\boldsymbol{B}$), *and we write* $\boldsymbol{B} = \boldsymbol{B}_1\boldsymbol{B}_2$ *to mean that $\boldsymbol{B}$ is the beta-binder given by the juxtaposition of $\boldsymbol{B}_1$ and $\boldsymbol{B}_2$.*

Also, the metavariables $\boldsymbol{B}^*, \boldsymbol{B}_1^*, \boldsymbol{B}_2^*, \ldots$ *stay for either a well-formed beta-binder or the empty string. The above notation for the subject function and for juxtaposition is extended to these metavariables in the natural way.*

Beta-processes (ranged over by $B, B_1, \ldots, B', \ldots$) are generated by the following grammar:

$$B ::= \mathsf{Nil} \ \bigg| \ \boldsymbol{B}[P] \ \bigg| \ B \parallel B$$

Nil denotes the deadlocked box and is the neutral element of the parallel composition of beta-processes, written $B \parallel B$. But for Nil, the simplest form of beta-process is given by a pi-process encapsulated into a composite beta-binder ($\boldsymbol{B}[P]$).

To any beta-process consisting of n parallel components, it corresponds a simple graphical notation given by n distinct boxes. Each box contains a pi-process and has as many sites (hidden or not) as the number of elementary beta-binders in the composite binder. The relative position of sites along the perimeter of the box is irrelevant.

Beta-processes are given an operational reduction semantics that makes use of both a structural congruence over beta-processes and a structural congruence over pi-processes. We overload the same symbol to denote both congruences, and let the context disambiguate the intended relation.

Definition 3. Structural congruence *over pi-processes, denoted* $\equiv$, *is the smallest relation which satisfies the following laws.*

- $P_1 \equiv P_2$ *provided* P_1 *is an* α-*converse of* P_2
- $P_1 \mid (P_2 \mid P_3) \equiv (P_1 \mid P_2) \mid P_3$, $P_1 \mid P_2 \equiv P_2 \mid P_1$, $P \mid nil \equiv P$
- $\nu z \, \nu w \, P \equiv \nu w \, \nu z \, P$, $\nu z \, nil \equiv nil$,
  $\nu y \, (P_1 \mid P_2) \equiv P_1 \mid \nu y \, P_2$ *provided* $y \notin fn(P_1)$
- $!P \equiv P \mid !P$

Structural congruence *over beta-processes, denoted* $\equiv$, *is the smallest relation which satisfies the laws listed below, where* $\hat{\beta}$ *is intended to range over* $\{\beta, \beta^h\}$.

- $\boldsymbol{B}[P_1] \equiv \boldsymbol{B}[P_2]$ *provided* $P_1 \equiv P_2$
- $B_1 \parallel (B_2 \parallel B_3) \equiv (B_1 \parallel B_2) \parallel B_3$, $B_1 \parallel B_2 \equiv B_2 \parallel B_1$, $B \parallel Nil \equiv B$
- $\boldsymbol{B}_1 \boldsymbol{B}_2[P] \equiv \boldsymbol{B}_2 \boldsymbol{B}_1[P]$
- $\boldsymbol{B}^* \hat{\beta}(x : \Gamma)[P] \equiv \boldsymbol{B}^* \hat{\beta}(y : \Gamma)[P\{y/x\}]$ *provided* y *fresh in* P, Γ *and* $y \notin sub(\boldsymbol{B}^*)$

The laws of structural congruence over pi-processes are the typical axioms of the π-calculus. The laws over beta-processes state, respectively, that (i) the structural congruence of pi-processes is reflected at the upper level as congruence of boxes; (ii) the parallel composition of beta-processes is a monoidal operation with neutral element Nil; (iii) the actual ordering of elementary beta-binders within a composite binder is irrelevant; (iv) the subject of elementary beta-binders is a placeholder that can be changed at any time under the proviso that name clashes are avoided and well-formedness of the composite binder is preserved.

The *reduction relation*, $\longrightarrow$, is the smallest relation over beta-processes obtained by applying the axioms and rules in Table 1.

The reduction relation describes the evolution within boxes (intra), as well as the interactions between boxes (inter), the dynamics of box interfaces (expose, hide, unhide), and the structural modification of boxes (join, split).

The rule intra lifts any 'reduction' of the enclosed pi-process to the level of the enclosing beta-process. Notice indeed that no reduction relation is defined over pi-processes.

The rule inter models interactions between boxes with complementary internal actions (input/output) over complementary sites (sites with non-disjoint types). Information flows from the box containing the pi-process which exhibits

Table 1. Axioms and rules for the reduction relation

(intra)
$$\frac{P \equiv \nu\tilde{u}\,(x(w).\,P_1 \mid \overline{x}\langle z\rangle.\,P_2 \mid P_3)}{\boldsymbol{B}[P] \longrightarrow \boldsymbol{B}\big[\nu\tilde{u}\,(P_1\{z\!/\!w\} \mid P_2 \mid P_3)\big]}$$

(inter)
$$\frac{P \equiv \nu\tilde{u}\,(x(w).\,P_1 \mid P_2) \qquad\qquad Q \equiv \nu\tilde{v}\,(\overline{y}\langle z\rangle.\,Q_1 \mid Q_2)}{\beta(x:\Gamma)\,\boldsymbol{B}_1^*[P] \parallel \beta(y:\Delta)\,\boldsymbol{B}_2^*[Q] \longrightarrow \beta(x:\Gamma)\,\boldsymbol{B}_1^*[P'] \parallel \beta(y:\Delta)\,\boldsymbol{B}_2^*[Q']}$$

where $P' = \nu\tilde{u}\,(P_1\{z\!/\!w\} \mid P_2)$ and $Q' = \nu\tilde{v}\,(Q_1 \mid Q_2)$

provided $\Gamma \cap \Delta \neq \emptyset$ and $x, z \notin \tilde{u}$ and $y, z \notin \tilde{v}$

(expose)
$$\frac{P \equiv \nu\tilde{u}\,(\mathsf{expose}(x,\,\Gamma).\,P_1 \mid P_2)}{\boldsymbol{B}[P] \longrightarrow \boldsymbol{B}\,\beta(y:\Gamma)\big[\nu\tilde{u}\,(P_1\{y\!/\!x\} \mid P_2)\big]}$$

provided $y \notin \tilde{u}$, $y \notin \mathsf{sub}(\boldsymbol{B})$ and $y \notin \Gamma$

(hide)
$$\frac{P \equiv \nu\tilde{u}\,(\mathsf{hide}(x).\,P_1 \mid P_2)}{\boldsymbol{B}^*\,\beta(x:\Gamma)\big[P\big] \longrightarrow \boldsymbol{B}^*\,\beta^h(x:\Gamma)\big[\nu\tilde{u}\,(P_1 \mid P_2)\big]}$$

provided $x \notin \tilde{u}$

(unhide)
$$\frac{P \equiv \nu\tilde{u}\,(\mathsf{unhide}(x).\,P_1 \mid P_2)}{\boldsymbol{B}^*\,\beta^h(x:\Gamma)\big[P\big] \longrightarrow \boldsymbol{B}^*\,\beta(x:\Gamma)\big[\nu\tilde{u}\,(P_1 \mid P_2)\big]}$$

provided $x \notin \tilde{u}$

(join) $\boldsymbol{B}_1[P_1] \parallel \boldsymbol{B}_2[P_2] \longrightarrow \boldsymbol{B}[P_1\sigma_1 \mid P_2\sigma_2]$

provided that f_{join} is defined at $(\boldsymbol{B}_1, \boldsymbol{B}_2, P_1, P_2)$

and with $f_{join}(\boldsymbol{B}_1, \boldsymbol{B}_2, P_1, P_2) = (\boldsymbol{B}, \sigma_1, \sigma_2)$

(split) $\boldsymbol{B}[P_1 \mid P_2] \longrightarrow \boldsymbol{B}_1[P_1\sigma_1] \parallel \boldsymbol{B}_2[P_2\sigma_2]$

provided that f_{split} is defined at $(\boldsymbol{B}, P_1, P_2)$

and with $f_{split}(\boldsymbol{B}, P_1, P_2) = (\boldsymbol{B}_1, \boldsymbol{B}_2, \sigma_1, \sigma_2)$

(redex)
$$\frac{B \longrightarrow B'}{B \parallel B'' \longrightarrow B' \parallel B''}$$
(struct)
$$\frac{B_1 \equiv B_1' \qquad B_1' \longrightarrow B_2}{B_1 \longrightarrow B_2}$$

---

the output prefix to the box enclosing the pi-process which is ready to perform the input action.

The rules expose, hide, and unhide correspond to an unguarded occurrence of the homonymous prefix in the internal pi-process and allow the dynamic modification of external interfaces. The rule expose causes the addition of an extra site with the declared type. The name x used in expose(x, Γ) is a placeholder which can be renamed to meet the requirement of well-formedness of the enclosing beta-binder. The rules hide and unhide force the specified site to become hidden or unhidden, respectively. Both these two rules cannot be applied if the interested site does not occur unhidden, respectively hidden, in the enclosing interface.

The axiom join models the merge of boxes. The rule, being parametric w.r.t. the function f_{join}, is more precisely an axiom schema. The function f_{join} determines the actual interface of the beta-process resulting from the aggregation of boxes, as well as possible renamings of the enclosed pi-processes via the substitutions σ_1 and σ_2. It is intended that as many different instances of f_{join} (and hence of the join axiom) can be defined as it is needed to model the system at hand.

The axiom split formalizes the splitting of a box in two parts, each of them taking away a subcomponent of the content of the original box. Analogously to join, the rule split is an axiom schema that depends on the specific definition of the function f_{split}. Analogously to the case of the join axiom, many instances of split can live together in the same formal system.

The rules redex and struct are typical rules of any reduction semantics. They are meant, respectively, to interpret the reduction of a subcomponent as a reduction of the global system, and to infer a reduction after a proper structural shuffling of the relevant processes.

4   An Example: Interaction Between P53 and Mdm2

In this section we illustrate how Beta-binders can be used to model Kohn MIMs. Here we consider the simple, and yet representative, sub-map of the mammalian cell cycle reported in Figure 2. The same modelling technique could be extended to render the entire map, nonetheless the small subset we analyze is sufficient to show the main features of the approach.

First, a beta-process is let to correspond to each biological entity drawn in the Kohn map (circles and rectangles, but the degradation symbol).

Each beta-process may have sites for binding to other elements, sites for activation and inhibition, and sites for covalent modifications. In particular, one site is associated to a given box for each kind of interaction represented on each line entering or leaving the corresponding biological element. For example, consider in Figure 2 the p53 element and the Ps15 and the Ps37 lines, that stay for serine15 and serine37, respectively. The beta-process corresponding to p53 is added with two sites for the first line (to render both phosphorylation and dephosphorylation of the bond), and one single site for the Ps37 line (for phosphorylation only).

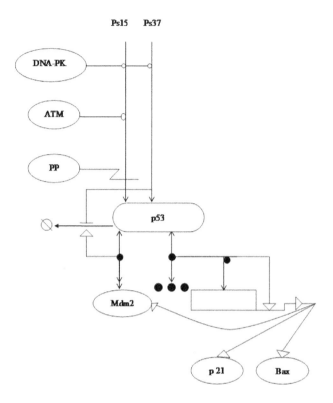

Fig. 2. Kohn map relative to p53 interactions

Finally, the fine-grain activity of the various proteins and genes are translated into the pi-processes enclosed into the various beta-processes.

Relatively to the map in Figure2, we get the formal system below.

$$S = P53 \parallel P53 \parallel P53 \parallel P53 \parallel$$
$$MDM2 \parallel ATM \parallel DNAPK \parallel PROM \parallel BAX \parallel P21 \parallel PP$$

The names of the parallel components of S, even if completely capitalized, have been chosen by analogy with the biological components they are meant to represent. The main role of each of them is as it follows.

- The four copies of $P53$ are used to allow the reprentation of the transformation of p53 from monomer to tetramer.
- $MDM2$ plays the role of the main negative regulator of p53.
- ATM and $DNAPK$ represent kinases able to phosphorylate p53 at different sites.
- $PROM$, the promoter, is the main actor of the activation of the p53-related genes. The activation is stimulated when p53 is in its tetramer state.
- BAX and $P21$ stay for p53-activated genes, the first is involved in cell apoptosis, and the latter in growth arrest.
- PP models the phosphatase that dephosphorylates p53 after DNA repair.

$$P53 \quad = \beta(x : \{Tps15\}; y : \{Tps37\}; u : \{Tmono\}; z : \{TNterm\})$$
$$\beta^h(x' : \{Tds15\})$$
$$[\,!\,x.\ \mathsf{hide}(x).\ \mathsf{unhide}(x').\,x'.\ \mathsf{hide}(x').\ \mathsf{unhide}(x)\mid y.\ \mathsf{hide}(y)\,]$$

$$DNAPK = \beta(x : \{Tps15\}; y : \{Tps37\})\,\big[!\,\overline{x}\mid\overline{y}\big]$$

$$ATM \quad = \beta(x : \{Tps15\})\,\big[!\,\overline{x}\big]$$

$$PP \quad = \beta(x : \{Tds15\})\,\big[!\,\overline{x}\big]$$

$$PROM \quad = \beta^h(x : \{Tpromote, Ttetra\})\,\big[!\,\overline{x}\big]$$

$$BAX \quad = \beta(z : \{Tpromote\})\,\big[z.\ \mathsf{hide}(z)\,\big]$$

$$P21 \quad = \beta(z : \{Tpromote\})\,\big[z.\ \mathsf{hide}(z)\,\big]$$

$$MDM2 \quad = \beta(z : \{Tpromote\})\,\beta^h(w : \{TNterm\})\,\big[z.\ \mathsf{hide}(z).\ \mathsf{unhide}(w)\,\big]$$

Fig. 3. Specification of the components of system S

The actual specification of the parallel components of system S is reported in Figure 3. Also, to complete the Beta-binders representation of the analyzed p53 interactions, suitable instances of f_{join} are defined to model the relevant bindings drawn in Figure 2, namely the homopolymer formation, the p53-promoter binding, and the Mdm2-p53 binding.

In Figure 3, to improve on readability, we use the following conventions. Whenever the actual names of action parameters are not relevant to the running explanation, inputs and outputs over the name x are simply denoted by x and $\overline{x}$, respectively. Also, the composite beta-binder $\beta(x_1 : \Gamma_1)\ldots\beta(x_j : \Gamma_j)\,\beta^h(y_1 : \Gamma_1)\ldots\beta^h(y_k : \Gamma_k)$ is shortly denoted by $\beta(x_1 : \Gamma_1; \ldots; x_j : \Gamma_j)\,\beta^h(y_1 : \Gamma_1; \ldots; y_k : \Gamma_k)$. Eventually, trailing '. nil's are sistematically omitted. In this respect notice that here we are just rendering a limited subset of the whole biological picture reported by Kohn in [6]. Then, for instance for the p53-related genes, we are only concerned to represent the way they are activated by the promoter, and any other further behaviour of these genes is abstracted into a deadlock.

The possible evolutions of system S reflect the biological dynamics of the analyzed MIM as it is illustrated below.

Phosphorylation of p53 is rendered as possible interactions at two sites of the $P53$ process ($x : \{Tps15\}$, and $y : \{Tps37\}$). In particular, inter reductions may be applied, through x and y, between $P53$ and either of ATM and $DNAPK$, namely the two processes that play the corresponding kinases. In either case, after phosphorylation the site involved in the reduction is hidden. Consider for instance the pi-subprocesses '$!\,x.\ \mathsf{hide}(x).\ \ldots$' of $P53$ in combination with the subprocess '$!\,\overline{x}$' of $DNAPK$.

Recall by Figure 2, that the phosphorylation of p53 at serine15 may be disrupted by a dephosphorylation due to the phosphatase PP. For this reason, the $P53$ specification is such that, after the possible inter-communication over x and the subsequent hiding of the site, the dephosphorylation site $x' : \{Tds15\}$ is unhidden and the process becomes ready to interact with the external world through x'. Notice also that the replication operator is used in the pi-processes enclosed in $P53$, $DNAPK$, ATM, and PP to represent the fact that phosphorylation and dephosphorylation of p53 at serine15 may keep alternating.

The formation of complexes involves the application of join reductions. Four distinct instances of f_{join} are used to represent the various kinds of binding. Only three of them are actually drawn in Figure 2. Here we use four instances of f_{join} to model the homopolymer formation in two steps (two monomers join to make a dimer, and then two dimers join to make a tetramer).

The actual definition of the relevant f_{join} functions is given in Figure 4, where σ_{id} stays for the identity substitution, and $\perp$ for undefinedness. The various instances of f_{join} can be commented upon as it follows.

1. The first instance, f_{join_1}, is used to get the p53 dimer formation. Two beta-processes exposing a $Tmono$-typed site can reduce to a unique box, with one $Tdimer$-typed site replacing the two $Tmono$ sites and with doubled copies of all the other $P53$ sites.

   The definition of f_{join_1} is such that the possible external interactions through the $Tmono$-typed sites are transfered, after the join, to the $Tdimer$-typed site (cf the actual instantiations of σ_1, σ_2). Also notice that the structural rule of the operational semantics allows the subjects of beta-binders to be refreshed so to meet the well-formedness requirement in the definition of f_{join_1}.

2. The tetramer formation is obtained by applying a join reduction under the definition of f_{join_2}. Analogously to the case of f_{join_1}, here a $Ttetra$-typed site substitutes the two original $Tdimer$-typed sites.

3. The third instance of f_{join} is the one rendering the binding between p53 in its tetramer state and the promoter element. The box for the protein exhibits a $Ttetra$-typed site. The promoter is a box with a hidden site $x : \{Tpromote, Ttetra, \ldots\}$.

   The join reduction under f_{join_3} unhides the x site for the transcriptional activation of the p53-related genes. Indeed, from the specifications in Figure 3, one can observe that a beta-process offering replicated output actions over a $Tpromote$-typed site can have inter reactions with each of BAX, $P21$, and $MDM2$. Also notice that the activation of $MDM2$ causes a $TNterm$-typed site to be unhidden on the corresponding box. This reflects the fact that, when activated by the promoter element, Mdm2 can bind to the N-terminus of p53.

4. The latest instance of f_{join} is meant to describe both the binding between Mdm2 and p53 in a non-phosphorylated state and the subsequent degradation of p53. The join reduction driven by f_{join_4} results in a beta-process with the same sites as those of $MDM2$. So, the degradation of p53 is

rendered by the fact that further external interactions of the original $P53$ process are actually impossible after the two boxes have been joined together.

A possible evolution of the system is reported in Figure 5. It illustrates the phosphorylation of p53 at the serine15 site by the ATM kinase. Beta-processes not directly involved in that specific interaction, just as the irrelevant sites, are omitted from the picture.

$f_{join_1} = \lambda \boldsymbol{B}_1 \boldsymbol{B}_2 P_1 P_2.$
  if   ($\boldsymbol{B}_1 \boldsymbol{B}_2$ is well-formed and
    $\boldsymbol{B}_1 = \boldsymbol{B}_1^* \beta(x : \{Tmono\})$ and $\boldsymbol{B}_2 = \boldsymbol{B}_2^* \beta(y : \{Tmono\}))$
  then $(\beta(y : \{Tdimer\}) \boldsymbol{B}_1^* \boldsymbol{B}_2^*, \{y/x\}, \sigma_{id})$
  else $\bot$

$f_{join_2} = \lambda \boldsymbol{B}_1 \boldsymbol{B}_2 P_1 P_2.$
  if   ($\boldsymbol{B}_1 \boldsymbol{B}_2$ is well-formed and
    $\boldsymbol{B}_1 = \boldsymbol{B}_1^* \beta(x : \{Tdimer\})$ and $\boldsymbol{B}_2 = \boldsymbol{B}_2^* \beta(y : \{Tdimer\}))$
  then $(\beta(y : \{Ttetra\}) \boldsymbol{B}_1^* \boldsymbol{B}_2^*, \{y/x\}, \sigma_{id})$
  else $\bot$

$f_{join_3} = \lambda \boldsymbol{B}_1 \boldsymbol{B}_2 P_1 P_2.$
  if   ($\boldsymbol{B}_1 \boldsymbol{B}_2$ is well-formed and
    $\boldsymbol{B}_1 = \boldsymbol{B}_1^* \beta^h(x : \{Tpromote\} \cup \Delta_1)$ and
    $\boldsymbol{B}_2 = \boldsymbol{B}_2^* \beta(y : \Delta_2)$ and $Ttetra \in \Delta_1 \cap \Delta_2)$
  then $(\boldsymbol{B}_1^* \beta(x : \{Tpromote\} \cup \Delta_1) \boldsymbol{B}_2, \sigma_{id}, \sigma_{id})$
  else $\bot$

$f_{join_4} = \lambda \boldsymbol{B}_1 \boldsymbol{B}_2 P_1 P_2.$
  if   ($\boldsymbol{B}_1 \boldsymbol{B}_2$ is well-formed and
    $\boldsymbol{B}_1 = \boldsymbol{B}_1^* \beta(x : \{Tps15\}; y : \{Tps37\}; z : \Delta_1)$ and
    $\boldsymbol{B}_2 = \boldsymbol{B}_2^* \beta(y : \Delta_2)$ and $TNterm \in \Delta_1 \cap \Delta_2)$
  then $(\boldsymbol{B}_2, \sigma_{id}, \sigma_{id})$
  else $\bot$

Fig. 4. Instances of f_{join}

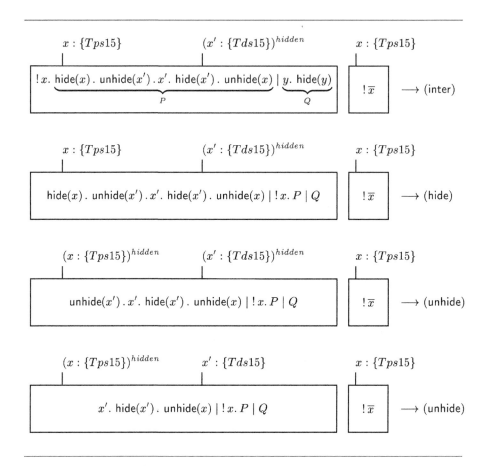

Fig. 5. Phosphorylation of $p53$ at serine15 by ATM – only $(P53 \parallel ATM)$ is shown

5   Concluding Remarks

A small subset of the Kohn MIM for the mammalian cell cycle was modeled using a formalism that, like process calculi, allows the description of the dynamic evolution of complex systems, and can be the basis for the development of automatic tools for inferring their behaviour.

Although the chosen map is just a small subset of the one appeared in [6], the translation we proposed deals with a representative collection of the biological reactions and interactions statically shown in MIMs.

Other formalisms have already been shown to be suitable to translate MIMs. Here we recall the *bio-calculus* [10], and the *core formal language for molecular biology* [4]. The first of them was intentionally meant to offer a language as close as possible to the conventional expressions used by biologists. In this respect, the bio-calculus describes a biological phenomenon as a set of expressions that are associated with diagrams and also include information for computational analysis. From the operational point of view, these expressions are interpreted as a sort of rewriting rules over an initial status set that is represented as $component\_1$ $| \ldots |$ $component\_n$. For instance, given the initial status set $proteinA \mid kinaseC$, the binding of the two components is modelled by a rule like the following:

$$proteinA \quad | \quad kinaseC \; \Rightarrow \; proteinA\_kinaseC \,. \tag{1}$$

The formalism introduced in [4] is based on rewriting rules, too. The approach, however, is substantially different from the one taken in bio-calculus. The representation of each single component is coupled with the explicit representation of its state (given by the occupancy of the sites of the component itself). So the rewriting rules adopted in the core formal language can also capture modifications at the level of sites. For instance, a notation like $proteinA\langle free, busy\rangle$ is used to represent a component with two sites, one free and the other already taken. Then in the core formal language the rule corresponding to (1) would give evidence to the state of the single sites after binding. Indeed it would be something like the following:

$proteinA \; \langle free, busy\rangle \mid kinaseC \; \langle free\rangle \Rightarrow proteinA \; \langle busy, busy\rangle \mid kinaseC \; \langle busy\rangle.$

The operational underpinning of Beta-binders is quite different from those of either the bio-calculus or the core formal language. A full comparison between the (translations of MIMs in the) three formalisms would require a long investigation and is far beyond the scope of the present paper. We can just briefly comment on this issue by mentioning that the distinctive feature of the process algebra approach is the compositionality it ensures. Specifying a process amounts to assemble together separate parallel components whose representations could be defined in isolation. The behaviour of the process is then determined after inference rules which are *a priori* defined. The situation is quite different with rewriting systems. There, analyzing bigger systems typically corresponds to adding new rules, and also modifying some of those already stated.

References

1. B. Alberts, A. Johnson, J. Lewis, M. Raff, K. Roberts, and P. Walter. *Molecular biology of the cell (IV ed.)*. Garland science, 2002.
2. L. Cardelli. Membrane interactions. In *BioConcur '03, Workshop on Concurrent Models in Molecular Biology*, 2003.
3. V. Danos and J. Krivine. Formal molecular biology done in CCS-R. In *BioConcur '03, Workshop on Concurrent Models in Molecular Biology*, 2003.
4. V. Danos and C. Laneve. Core formal molecular biology. In P. Degano, editor, *Proc. 12th European Symposium on Programming (ESOP '03)*, volume 2618 of *Lecture Notes in Computer Science*. Springer, 2003.
5. M.L. Guerriero. Modelli formali per la rappresentazione di sistemi biologici dinamici. Laurea Thesis, University of Trento, Dep. Informatics and Telecommunications (In Italian), 2004.

6. K.W. Kohn. Molecular interaction map of the mammalian cell cycle contro and DNA repair systems. *Molecular Biology of the Cell*, (10):2703–2734, 1999.
7. D. Lane. Surfing the p53 network. *Nature*, 408:307–10, 2000.
8. D. Lane, B. Vogelstein, and A.J. Levine. P53, the guardian of the genome. *Nature*, 358:15–16, 1992.
9. R. Milner. *Communicating and mobile systems: the π-calculus*. Cambridge Universtity Press, 1999.
10. M. Nagasaki, S. Onami, S. Miyano, and H. Kitano. Bio-calculus: Its concept and molecular interaction. In K. Asai, S. Miyano, and T. Takagi, editors, *Genome Informatics 1999*, volume 10, pages 133–143. Universal Academy Press, 1999.
11. C. Priami and P. Quaglia. Beta binders for biological interactions. In *Proc. CMSB '04*, Lecture Notes in Bioinformatics. Springer, 2004. To appear.
12. C. Priami and P. Quaglia. Operational patterns in Beta-binders. *Transactions on Computational Systems Biology*, 2004. To appear.
13. C. Priami, A. Regev, W. Silverman, and E. Shapiro. Application of a stochastic name-passing calculus to representation and simulation of molecular processes. *Information Processing Letters*, 80(1):25–31, 2001.
14. A. Regev, E.M. Panina, W. Silverman, L. Cardelli, and E. Shapiro. Bioambients: An abstraction for biological compartments. *Theoretical Computer Science*, 2004. To appear.
15. A. Regev, W. Silverman, and E. Shapiro. Representation and simulation of biochemical processes using the pi-calculus process algebra. In *Proc. of the Pacific Symposium on Biocomputing (PSB '01)*, volume 6, pages 459–470. World Scientific Press, 2001.
16. D. Sangiorgi and D. D. Walker. *The π-calculus: a Theory of Mobile Processes*. Cambridge Universtity Press, 2001.

Multidisciplinary Investigation into Adult Stem Cell Behavior

Mark d'Inverno[1] and Jane Prophet[2]

[1] Cavendish School of Computer Science,
University of Westminster,
115, New Cavendish Street, London W1M 8JS
dinverm@wmin.ac.uk
[2] University of Westminster, 70, Great Portland Street,
London W1W 7UW
jane@carte.org.uk

Abstract. We are part of a multi-disciplinary team investigating new understandings of stem cell behaviour, and specifically looking novel ideas about how adult stem cells behave and organise themselves in the human body. We have used different methods and mechanisms for investigating and interpreting these new ideas from medical science. These have included mathematical modelling, simulation and visualisation as well as a series of art pieces that have resulted from looking at the overall nature of our combined multi-disciplinary attempt to investigate new theories of biological organisation. In this paper we look at several issues relating to our project. First, we provide reasons for why formal models and simulations are needed to explore this growing area of research. Is there an argument to suggest that we need simulations as a way in to understanding various properties of stem cells that are observed in wet lab experiments? Next, an introduction on current theoretical models of stem cells is presented along with an outline of some of the problems and limitations of these approaches. We then provide an overview of our agent-based model of stem cells, and discuss strategies for implementing this model as a simulation and its subsequent visualisation. Then we discuss the role of the artist (the second author) in developing our model and simulation and the influence of the artwork/exhibition on the other members in our team. Our premise is that artists can conceptualise scientific theories without the standard discipline-specific constraints, and thereby potentially influence the development of scientific theories, their mathematical formulation; and their associated aesthetics. Finally, we argue that for the field of bioinformatics to move forward in a holistic and cohesive manner more multi-disciplinary efforts such as ours would be of significant benefit to this research area [20]. This paper might be viewed as an advert for the benefits of multi-disciplinary teams in understanding new experimental data in medicine and biology.

Keywords: Novel tools applied to biological systems, self-organizing, self-repairing and self-replicating systems, new technologies and methods, cellular automata, art, interdisciplinary research and collaboration.

C. Priami et al. (Eds.): Trans. on Comput. Syst. Biol. III, LNBI 3737, pp. 49–64, 2005.
© Springer-Verlag Berlin Heidelberg 2005

1   Introduction

Recent experimental evidence has suggested novel ways in which stem cells behave. The standard model, where a stem cell becomes increasingly differentiated over time, along a well-defined cell-lineage and eventually becomes a fully functional cell have been challenged by many researchers including one of our collaborators on a project entitled Cell [13,31,29]. Several years ago, new theories were proposed by our collaborator and others that challenged the prevailing view. Because of new experimental data it was suggested that stem cell fate is both *reversible* (cells can become less differentiated and more like stem cells) and is *plastic* (cells can *jump* from one lineage to another).

Whilst there have been attempts to predict stem cell behaviour in terms of the internal state of a cell, or the environment in which it is situated, the prevailing predominant view (as in our own team) that both the internal state of the cell, and the current state of the microenvironment, are critical in determining how stem cells behave. Moreover, models will should encompass the ability of cells to behave stochastically.

It became evident to us, coming from a background in agent-based systems and Alife, that an appropriate way in which to understand this new theory was as a complex system of interacting stem cell agents where global qualities emerged as a result of local behaviour determined by local environmental conditions. The behaviour of cells is determined solely by its own internal state and the local state of the environment. The agent is the ideal way to understand the interplay of internal and environmental factors on the behaviour of stem cells in general. Our approach has been to model stem cells as agents in a self-organising system that does not rely on global data structures, information about the system or statistical devices such as probability density functions [30,5].

There have been two strands to our work. First we have built a model and simulation in collaboration with a leading stem cell researcher of new principles of stem cell behaviour and organisation using agent-based modelling techniques. Second we have reformulated existing models and simulations within our agent framework thus allowing us to evaluate, compare, and incorporate techniques from other models. In this paper we will provide details of our multi-disciplinary project which has been concerned with investigating this new theory and the possible ramifications of this new mode of understanding. The methods we have used to investigate this theory and the outcomes from it includes those listed below.

- A *formal model* which represents a definitive abstract functional model that embodies this new theory of stem cell fate and organisation. This helps to catalyse a common conceptual framework for the terms and artefacts of our work.
- Reformulation of *existing models of stem cell behaviours*. This is done for a number of reasons. First to ensure that we are building a unifying framework for modeling stem cell behaviour and that our model is sufficiently expressive to encompass existing models, second to compare and evaluate the modelling

techniques of existing models, and third to incorporate existing concepts, structures and processes into our own agent-based model.

- Our own formal representation and simulation of *existing models of stem cell behaviours*. This is done for a number of reasons. First to provide a unifying framework for modeling stem cell behaviour, second to ensure that our model is sufficiently expressive to encompass existing models, thirdly to compare and evaluate these existing models, and fourth to incorporate various concepts and structures into our own models.

- A *simulation* that embodies the formal model using a functional programming language to build the key components (cells). Whilst there are several existing mathematical models of stem cells, there are very few that have been simulated. Without simulation, properties of the resulting model must be proved [1], and the models are extremely simple and lack sufficient biological realism. If we wish to study how cells interacting at a local level produce observable global phenomenon, such as the ability of stem cells to generate requisite numbers of functioning cells after disease or injury, simulation is vital, since it would not be possible to prove general properties of a more sophisticated model in practice.

- Several *visualisations* have been designed and implemented. We see the process of visualisation to be distinct, although related, to the simulation itself. Visualisation is the process of making observable to a user some of the events and processes that are occurring in the simulation.

- An *art exhibition* that not only reflects on the model but on the process of collectively producing the artefacts listed above. The first artwork, *Staining Space*, reflected on issues of how we might, collectively, reconceptualise science. Attempting such a reconceptualisation, however minimal, recognises ways in which art can influence scientific research in a quantitative as well as a qualitative way.

- A *generic model for interdisciplinary collaboration*. Our collaborative model is one of convergence where experts from different disciplines come together to discuss a range of related topics. This is followed by sustained enquiry by the whole group. Subsequently, the group diverges to produce a range of outcomes each of which is peer-reviewed from within the appropriate different discipline. Some outcomes are co-produced by individuals who would not usually author together. This results in what has been termed *transvergence* by Marcos Novak [18]- the individuals involved in the collaboration, and their disciplines, are transformed through the process.

In this paper we provide an overview of our project, and of the various outputs that have been produced a result.

1.1   Overview

In this paper we will first provide a background into the role and importance of modelling and simulation in stem cell research and provide a couple of examples of work done to date. We will then provide some details of the model and our view that adopting an agent-based/complex-system approach to modelling stem cells is appropriate. We will also provide some discussion as to the visualisation in general. In the next section switch to describing the art exhibition called staining space that resulted from this collaboration and the impact of this work on members of the team.

2   Background

2.1   Formal Modelling of Stem Cell Systems

Many laboratory experiments involve looking at stem cell systems using a 2-dimensional slice of dead material, removed from the original living system using a scalpel or syringe, and often stained using various dyes and colours. In our view, such experiments cannot possibly evaluate and understand stem cells as a complex living self-regulating dynamic system. Mathematical modelling and subsequent simulation is important means to investigate stem cell properties in a systematic manner. Specifically, how individual stem cells behave in certain local environments, and how the sum of all these behaviours may lead to emergent properties such as maintaining a constant number of fully determined functional cells even after massive perturbations.

Although attempting to understand the organisation and behaviour of stem cells within the adult human body as a complex system has not received a large amount of attention there have been notable attempts to build formal theoretical models to investigate specific stem cell properties. Indeed, over the last year or so there has been a noticeable climate change in this respect, and there is now a growing awareness of the need to use mathematical modelling and computer simulation to understand the processes and behaviours of stem cells in the body. Some reasons have been pointed out by others [26], most comprehensively in a recent excellent survey [32] of mathematical techniques for predicting behaviours of stem cells. We summarise some of the key points here.

1. In the adult body stem cells cannot be distinguished morphologically from other primitive non-differentiated cell types.
2. Extracting stem cells from an embryo means sacrificing it, posing serious ethical difficulties.
3. There is no way to determine whether any individual isolated cell is a stem cell, or, to be able to model what its potential behaviour might be. It is not possible to make any definite statements about this cell. At best it can be tracked and its behaviour observed though clearly any particular behaviour is simply one of many possibly paths. The notion of a stem cell refers to the wide-ranging set of potential behaviours that it might have, and these are influenced by internal, environmental, and stochastic processes.
4. The number of possible interactions and behaviours of a large number of stem cells makes the system an extremely complex (in all the senses described above) one. Theoretical simplifications are therefore key to understanding fundamental properties [27].

There is thus a need for new theoretical frameworks and models which can be directly mapped to a computer simulation and which look at the dynamic self-organisation of stem cells. In our work we have incorporated existing models into our framework to ensure that it is encompassing and to demonstrate its general utility. We specifically wish to incorporate new ideas of plasticity where stem and reversibility which have been proposed by our collaborators and others in recent years to explain new experimental discoveries. There have been a few attempts to consider modelling stem cell systems that specifically consider the reversibility of stem cells. (To be more specific about what is understood by reversibility it is where any daughter of a stem cell has an affinity or property that is more stemlike than its parent.) Most notably is the innovative work or Roeder and Loeffler [15,27] who have specifically looked at these issues, and which we will consider in more detail later in the paper. The most recent attempt that we are aware of looking at the issue of reversibility is the work of Kirkland [10].

3   CELL Project: An Agent-Based Approach to Modelling Stem Cells

We have built a formal model of stem cells using an agent-based approach inspired by previous work in this area [3,4,16,17]. The general properties that we list and outline here have been instrumental in guiding our work in modelling the society of stem cells.

- Agents are autonomous entities that seek to maximise some measure of usefulness (which could relate to a goal, motivation, utility) by responding to the local environment according to a set of rules that define their behaviour. We are not suggesting that goals need to be represented explicitly for cells, only that the agent-based metaphor of *entities having purpose* may be an appropriate abstraction mechanism for modelling state and behaviours.
- Individual cell agents are not aware either of the larger organisation, goals, or needs of the system. Moreover, they are not aware of the overall state of the system. The cell's behaviour is defined solely in terms of its state, its local environment and, possibly, its perception of the local environment.
- We allow for the agents' behaviour to be non-deterministic (stochastic) in keeping with the latest ideas in the modelling of stem cells [26]. In fact much of the non-determinism is based on agents attempting to perform some action (for example divide or move to some location) in a dynamic changing environment, where the effects of actions are not guaranteed. For example, when two cells attempt to move to the same location in a niche, it is not possible that both will be successful.
- Agents can perceive aspects of the environment they are in and can act so as to change the state of the environment. Critically in a complex system agents must affect the environment in such a way that the resulting environmental change can both be perceived by others and affect the future behaviour of others.

– Different cell agents will have different perceptual capabilities, and they will apply those perceptual capabilities in different ways, possibly non-deterministically. Moreover, they may be able to decide (perhaps non-deterministically) on which, if any, of the perceptions of their environment they choose to react to.
– If there are multiple cells that interact in an environment, so that the environment changes in such a way that the environment effects the behaviours of other cells and so on, then it should be clear that as numbers increase then it becomes impossible to predict the behaviour of the system using formal techniques alone and and the nature of the system can only be investigated by allowing the system to "run" as a computer simulation. We are specifically interested in ways to understand how global system properties - such as stem cells maintaining a constant number of determined cells - from simple behavioural rule for are stem cells. An approach well-established in several branched of agent-based design such as Artificial Life; exploiting the idea of computational emergence as defined by Cariani [2].

In order to help justify why agents are appropriate we look at two existing approaches. We believe that modelling stem cells as agents, which have internal state, perceptual capabilities, and that respond to their local environment in which they are situated provides us with more biological plausible and intuitive models than existing approaches. The existing approaches we outline here are the cellular automata approach of Agur et al., and the previously mentioned work or Roeder looking at the reversibility of cells, which uses probability density functions based on global system information in order to determine stem cell fate.

3.1 Cellular Automata Approach

Existing approaches to understanding general properties of stem cells are often based on cellular automata. One recent example which we have looked at in detail is the approach of Agur et al. [1]. In this work they model a stem cell niche (places in the human body where stem cells reside) in such a way that they can prove that it has the ability to maintain a reasonably fixed number of stem cells, produce a supply of mature (differentiated) cells, and remain capable of returning to this state even after very large perturbations that might occur through injury or disease. The behaviour of a cell is determined by both internal (intrinsic) factors (a local clock) and external (extrinsic) factors (the prevalence of stem cells nearby) and so in some respect this is very similar to the agent-based model we have outlined above; cells have different states, and act and change the state of the environment, while different cells have different perceptual capabilities and so on.

Essentially, stem cell behavior is determined by the number of its stem cell neighbors. This assumption is aimed at simply describing the fact that cytokines, secreted by cells into the micro-environment are capable of activating quiescent stem cells into proliferation and differentiation.

Each cell has an internal state (a counter), which determine stem cell proliferation, stem cell transition into differentiation, as well as the transit time of a differentiated cell before migrating to the peripheral blood.

The niche is modelled as a connected, locally finite undirected graph of nodes. Each node is either empty, or occupied by either a stem cell or a differentiated cell. A stem cell is able to perceive information from neighbouring locations. Cell division (called proliferation), is modelled as is determination (where stem cells become determined cells without division). Determined cells stay in the niche for a period and then eventually leave the niche to enter the bloodstream.

There are three constant values (N1, N2 and N3) that are used to reflect experimental observation. The first constant (N1) represents the time taken for a differentiated cell to leave the niche. The second (N2) represents the cycling phase of a stem cell; a certain number of ticks of the clock are needed before the cell is ready to consider dividing. Finally, the third (N3) represents the amount of time it takes for an empty space that is continuously neighboured by a stem cell, to be populated by a descendent from the neighbouring stem cell.

The rules of their model, expressed in English, are as follows.

1. Determined cells
   (a) If the internal clock has reached N1 then leave the niche. Reset local clock to 0.
   (b) If the internal clock has not yet reached N1 then increment it 1.
2. Rule for stem cells
   (a) If the counter at a stem cell location has reached N2, and all stems are neighbours, then become a differentiated cell. Reset the clock to 0.
   (b) If the counter of a stem cell is equal to N2 but not all the neighbours are stem cells then do nothing. Leave clock unchanged.
   (c) If the counter has not reached N2 then do nothing except increment clock. Rule for empty spaces.
3. Rules for empty nodes
   (a) If the counter at the empty node has reached N3 and there is a stem cell neighbour then introduce (give birth to) a stem cell in that location. Reset the clock.
   (b) If the counter at an empty grid has not reached N3 and there is a stem cell neighbour then increase clock.
   (c) If there are no stem cell neighbours at all then reset the clock to 0.

The next state of the system is a function of the state of the cell, and the state of the neighbouring cells. All locations are then updated simultaneously. The benefit of this approach is that formal proofs about system behaviour can be developed.

However, there are a number of issues with this model that we have highlighted by reformulating this model as an agent-based system. First, the authors clearly state that the agents are influenced by their internal state and the state of their local environment, but in the detail of the formal model, the definitions make use of the fact that empty grid locations of the *environment* also have counters. That is empty nodes "do computational work" which seems biologically counter-intuitive. Moreover, divisionis not modelled explicitly, rather cells appear magically into empty spaces.

In our work we have reformulated this work formal model and simulation to provide a more biologically intuitive model of the stem cell niche, and of cell division, but keeping the basic original properties of the CA system. However, there is a general limitation of the cellular automata approach is the lack of a satisfactory model of the physical, biological and chemical environment. In our model we are including such mechanisms as collision avoidance, mechanical forces (see [6]), movement, chemical diffusion, different perceptual sensitivities and dynamic environments, secretion and inhibition. Even if not impossible, it is not clear how this considerations would be achieved with the cellular automata approach.

This is not to say that we do not realise the limitations of any simulation approach. Part of the advantage of working in a multi-disciplinary approach is that all the participants have been challenged about the way they use methods to investigate natural phenomenon. As a result, we have become increasingly aware of the gaps between the biological system, new medical theories of how such systems might work, our formal representation of one of these theories, the simulation of this theory and of a general environment which includes some physical and chemical modelling, the visualisation of this simulation (what aspects of the simulation should we attempt to visualise and how) and finally the resulting perception by any observer. The final perception of the visualisation is therefore many steps removed from the biological system. In our work we have tried to reduce this gap as much as possible by reducing the gaps (or at least being explicit about where the gaps may arise) between the individual steps.

The key disadvantage of the agent-based approach to modelling stem cells as compared to the cellular automata approach is that it will not allow for any formal proof about overall system behaviour. The system will effectively become a complex system [30] where any particular non-deterministic choice, for example by a cell agent, may have huge ramifications for the system. In one sense this is a disadvantage but in another it is exactly what we are looking for as we wish to determine what small changes in an individual cell's state or behaviour might lead to system catastrophes such as leukemia for example [31].

3.2   A Model of Stem Cell Self-organisation

From a biological viewpoint the model of Agur et al. does not allow any reversibility or plasticity in the basic properties of cells. For example, once a cell has differentiated it cannot become a stem cell again (or, in a more continuous view, more plastic). Moreover, once a cell has left the niche, it cannot return. A recent example, an approach that uses a more sophisticated model and addresses these issues, is that of Markus Loeffer and Ingo Roeder at the University of Leipzig, who model hematopoietic stem cells using various (but limited) parameters including representing both the growth environment within the marrow (one particular stem cell niche) and the cycling status of the cell [15]. The ability of cells to both escape and re-enter the niche and to move between high and low niche affinities (referred to as within-tissue plasticity) is stochastically determined. The validity of their model is demonstrated by the fact that it produces results in global behavior of the system that exactly match experimental laboratory observations. The point is that the larger patterns of system organization emerge from these few simple rules governing variations in niche-affinity and coordinated changes in cell cycle.

We do not provide details of the model here as it is much more complex than the Agur model. Indeed it is the most sophisticated model we have yet encountered. However, through the process of reformulating this work in an agent-based model and extending it to produce an agent-based simulation, it allows us to investigate emergence due to the subtle changes in micro-environmental effects for each cell acting as autonomous agent.

The essential issues are that the physical micro-environment is not explicitly modelled, and so the behaviour of cells is not determined by their local environment but rather by using probability density functions that are governed by global information such as the number of cells in the system. By introducing chemical secretion and diffusion into our agent-based approach have been able to re-formulate this model and replace this density function, so that the we can model cells as agent responding autonomously to their local environment. We do not need to use global information in order to direct behaviour. Moreover, as our approach is much more fine grained than the original statistical equations and allows for a much greater degree of sophistication in the possibilities of understanding how self-organisation actually takes place in the adult human body.

We are currently considering how other models can be included within our model and simulation framework, by reformulating them as an agent-based system (eg [10] and considering how other agent-based models of cell interaction [8] might be applicable to our own.

4   From Formal Model to Simulation

There are several reasons why we used a formal model in our project in addition to the general use of the formal system analysis for understanding stem cells listed above. Some of these were previously identified in the first author's collaboration with Michael Luck [4].

1. We wish the simulation to reflect as accurately as possible *any* theory of stem cell organisation. The formal model (a mathematical model *plus* plain English) is understandable by both the medical researcher (Theise) and the rest of the team including those responsible for building the simulation.
2. In any new research topic, as with multi-agent systems some years ago, it is necessary to build a well-defined and precise vocabulary for the fundamental elements of a system. If this model is intuitive and accessible even to those without a formal training then it catalyses what we have referred to

as a *common conceptual framework*, where there is a shared understanding of the essential concepts and terms. This is even more critical in a multi-disciplinary team where words and terms, because of the disparate nature of the individual members' backgrounds, become much more contentious.

3. One of our intentions is to build a software tool that can be tailored to experiment with any new theory of stem cell behaviours. In this regard it is important that our model is general enough so that it can enable alternative theories and models to be formalised, compared, evaluated and subsequently simulated. It must provide a description of the common abstractions found within that class of models as well as a means of further refining these descriptions to detail particular models and systems. To this effect we have already begun to take some of the key theoretical models and cage them within our model. If we can represent other models in our framework then it has additional practical value as well as providing evidence to validate the modelling abstractions we have used.

4.1   Choice of Tools: Modelling Language

We have built the formal model of our stem cells using the state-based specification language Z[28]. Z is our language of choice because we have considerable experience of using it to specify a framework for multi-agent systems [4]. It is a technique that enables designs of systems to be developed formally, and allows for these to be systematically refined to an implementation (simulation). The language has other benefits that have made it especially suitable for this project: it is more *accessible* than many other formalisms since it is based on existing *elementary* components such as set theory and first order predicate calculus. This was valuable when working in a multi-disciplinary team. Moreover, it is an extremely expressive language, producing a consistent and unified account of a system, its state and its operation.

5   Visualisation

The relationships between observation and visualisation, and between visualisation and simulation became hot topics in our discussions. Firstly, we identified important cultural and/or discipline-related differences in our understanding of these terms and fields. For the artist there was no such thing as an objective truth. The subject position of each observer, and their act of observing, inflected or changed the meaning of that which was observed. The second author, coming from a background in fine art and photographic theory had been trained to interpret any pictorial representation (from painting through to documentary photograph) as both subjective and of carrying meanings that defied objectivity. The arts hold that colour, composition, cropping, lighting etc. all affect meaning. Interpretation is not a passive act; it is an active act of intervention. In the arts, visualisation is also a performative act: there is a social component to all images that has a bearing on people doing things together [11]. This was not the case for Theise who regularly used photographs of stained tissue slides to *signify* the truth of his claims, unproblematically using such representations as *proof* and *truth*. Visualisation in the laboratory differs from visualisation in the art studio, and aesthetics are important to both. In the medical laboratory representation is usually taken literally, leading to scientific illustration. In the art studio *representation* is a term and process framed by debates in cultural theory and numerous theories of representation (for example, an image, sound, object can signify something without actually sounding or looking anything like it [7]).

The first author's perspective was very different. First, he is concerned that the formal model captures the stem cell researcher's understanding as completely and correctly as possible, and that the simulation therefore investigates the properties and consequences of this model in a dynamic system. The mathematical model is therefore a truthful representation of Theise's theory. He originally believed that the visualisation could embody this truth, and that a visualisation could have a common effect (i.e. an understanding of Theise's theory) on an audience.

However, we found common ground in our interest in Heisenberg's Principle of Uncertainty [9], in particular Theise's contentious interpretation of the principle that takes it to mean that the very act of observing a phenomenon inevitably alters that phenomenon in some way. Theise applied this to the field of liver pathology in general and adult stem cell research in particular, noting that the act of observing cell tissue irrevocably alters the tissue. To examine tissue under the microscope necessitates a disruption of its biological processes, up to the point of killing the tissue. We can expand on the principle to reflect that it is the observer as much as the observed that determines what is accepted as reality.

Having developed artificial life artworks (such as TechnoSphere, 1995 and Swarm, 1996 [19]), the second author *understood* stem cells as entities in a complex system and emphasised that killing tissue to observe cells alters the phenomena, and only shows one moment in time rather than providing a window onto the phenomena of dynamic cell behaviour across time. The slicing process that produces tissue slides also reduces three-dimensional space to two dimensions, and the staining of the tissue alters its *meaning* once again.

By testing hypotheses about stem cell behaviour using such fragmentary techniques, the part (a moment in time and two-dimensional space) stands in for the whole. It is our shared view that one of the most off-putting aspects of contemporary scientific research is the way that the human body is regularly reduced to *building blocks* (the cell, the gene). In conversations Theise critiqued the concept that cells are the body's *building blocks*, pointing out that early observations of cell walls through the microscope shaped the language of biology and determined the bounded cell as a key *unit*, reinforcing a reductive model of the human body in medicine. By contrast Theise's theory challenges the paradigm of the progressively differentiating adult stem cells.

Visualisation and simulation are often confused, or conflated. By contrast, we see them as distinct but related. Our mathematical model is an abstraction of the biological process, and the resulting agent-based simulation takes this abstraction one step further from the actual behaviour of cells in the human body. The visualisations we produce from these simulations further abstract (or distract) us from the real biological function that they represent. When faced with a choice between a flat, two-dimensional and clearly abstract presentation of the cell simulation (See Image 1 [22]), versus a photorealistic rendering of the same simulation (See Image 2 [23]), artist and scientists were drawn apart. Whist the artist preferred the two dimensional version as it was impossible to view it as anything other than an abstraction or approximation, both Theise and the first author preferred the more realistic looking three-dimensional version. Theise's choice was influenced by the fact that the 3D version looked more believable and could be presented publicly to an audience who would *suspend their disbelief* and thereby buy into the theory that the images presented. The look of this version was also considered to be beautiful and therefore preferable. While the importance of beauty in such bioinformation may seem inconsequential, anecdotal evidence shows a correlation between the beauty of images within scientific papers and the prominence given to those papers in publications, with so-called *beautiful* scientific images of all kinds given space on the covers of key journals.

Most contemporary artists have been educated .to resist beautifying the graphics. As far as possible, they want the graphic look and feel to reflect the underlying software, to draw attention to essence of the idea or concept with as little frippery and decoration as possible. From this standpoint the first version of the simulation is a more satisfying outcome than the 3D version. The 3D version has been influenced by the aesthetics of medical illustration and its goal of explaining via precise observation of the appearance of things: this is at odds with our emphasis on the behaviour of things (in this case stem cells). So, the notion of a unified aesthetic to the work emerging from the collaboration is contentious. Like much conceptual art, the idea behind the visualisations (namely modelling the behaviour of stem cells) and the means of producing it (via interdisciplinary collaboration) are more important than the finished work or its (fixed) appearance:

> In conceptual art the idea or concept is the most important aspect of the work . . . all planning and decisions are made beforehand and the execution is a perfunctory affair. The idea becomes the machine that makes the art. [14]

Conceptual art can be defined as the "appreciation for a work of art because of its meaning, in which the presentation of shape colour and materials have no value with out the intentions of the work". [12] If conceptual art has an aesthetic then it is the dematerialisation of the art-object; the object only has value as a materialisation of the idea, not in and of itself. Mathematics and computing science can both operate without materiality and can describe the immaterial, which is one reason why there may be a mutual attraction between computer scientists, mathematicians and artists working conceptually using digital media.

6   Staining Space

Staining Space is inspired by our discussions about the way that visualisation operates in a scientific context, for example microscopic analysis of tissue samples in the medical laboratory, and in a metaphoric way in a cultural context. It also addresses how the two have a constant inter-play. The installation poses questions that arise out of the ambiguities around visualising information and complex processes. Ultimately it is asking in what ways we perceive visual evidence as *real*.

The installation is also inspired by the challenge of representing or alluding to that which we cannot see. Things remain un-seeable because we lack a technology capable of visually recording them, and/or because we are unable to comprehend or accept what we are looking at. In pathology, stains are applied to tissues to make visible artefacts and structures that are otherwise invisible. *Staining Space* uses materials and forms to suggest the invisible being revealed; of images and their meanings being complex, contentious and multi-layered.

Seeing objects at different scales affects our perception. The actual scale of cells is hard to imagine as it is beyond the range of the naked eye, but it is comprehended as part of our intimate body-space. In the installation, a large (6m x 4m) still image of the digital cell simulation (see image 3 [24]) is offset against the small 2D visualisation of the simulation on a 4.5 inch monitor (see image 4 [25]). Similarly, a large projected tree is offset against the tree form in the tank, and they compete for the audience's attention. Frequently, when we visualise the invisible, we are not aware of the simplifications or manipulations that take place. Magnification is one example, as is reducing the representation of three dimensions to two dimensions, or taking a still image from a sequence that makes up a moving image.

Neither the digital animation nor the tree form have a *real* scale. Both are derived from data. The digital animation represents a theory of the way that stem cells develop and divide to create other specialised cells. Their changing colours indicate the degree of specialism, or differentiation, that they attain with every generation. The tree form is also derived from data *grown* initially in the Cartesian grid of a virtual 3D drawing-board. This data, presented physically is then chemically *grown* through crystallisation. The crystal tree alludes to a different notion of time. We understand that it has been growing at a rate we cannot perceive; that initially the seeded crystal string branches were bare and that the visible crystals will continue to grow in our absence.

The collaboration, and specifically the exhibitions' part in it is more than "posturing the subtle ambiguities of [scientific] knowledge and its human implication" [21]. Art changes our visions of the world, and can influence scientific research in a quantitative as well as qualitative way.

7   Conclusions

In this paper we have discussed the multi-disciplinary project CELL that is investigating new theories of stem cell organisation in the adult human body.

Our main goal has been to develop a software tool that medical researchers can use to test and run new hypothesis about the nature of stem cell behaviours in general but as a result of working in our team a number of other outputs have also arisen including art exhibitions, mathematical models, a new agent-based approach to simulating stem cells and novel visualisation techniques, some of which we have discussed here.

Our contention has always been that the art and the science should be more than just open to each other's influence, but should actively contribute to each other. Such contributions may not be demonstrated in week-by-week exchanges, but the mutual impact becomes clear as the project advances and *Staining Space* has embedded some of those impacts in a work of art [21].

Moreover, we claim that this research has a massive impact on the practices of us all, and perhaps most significant of all, on the stem cell researcher who was one of the first to propose the radical new view of stem cell fate as being plastic, reversible and non-deterministic. Not only has it had an effect on how he conceptualises this theory but also on the way he conducts experiments in the lab as a result.

The process of engaging with scientific discovery or research can be seen as part of a wider artistic project, that of making sense of our changing place in the world. While this depends to an extent on increasing our understanding of scientific research processes and outcomes, it also requires a critical distance that embeds those facts and discoveries in a wider cultural context. Collaborating across disciplines is not always easy, but is driven by an interest in how we might, collectively, reconceptualise science and develop new understanding about how we function and who we are.

Acknowledgements

The team of collaborators in this project (entitled CELL) not only included Neil Theise but also the curator Peter Ride and the A-life programmer Rob Saunders both from the University of Westminster. We would also like to thank the helpful referees for their comments.

References

1. Z. Agur, Y. Daniel, and Y. Ginosar. The universal properties of stem cells as pinpointed by a simple discrete model. *Mathematical Biology*, 44:79–86, 2002.
2. P. Cariani. Emergence and artificial life. In C. Langton, C. Taylor, J. Farmer, and S. Rasmussen, editors, *Artificial Life II*, pages 775–797, 1991.
3. M. d'Inverno, D. Kinny, and M. Luck. Interaction protocols in Agentis. In *IC-MAS'98 Proceedings of the Third International Conference on Multi-Agent Systems*, pages 112–119. IEEE Computer Society, 1998.
4. M. d'Inverno and M. Luck. *Understanding Agent Systems (Second Edition)*. Springer, 2004.

5. M. d'Inverno, N. D. Theise, and J. Prophet. Mathematical modelling of stem cells: a complexity primer for the stem cell biologist. In Christopher Potten, Jim Watson, Robert Clarke, and Andrew Renehan, editors, *Tissue Stem Cells: Biology and Applications*. Marcel Dekker, to appear, 2005.

6. M. Eastwood, V. Mudera, D. McGrouther, and R. Brown. Effect of mechanical loading on fibroblast populated collagen lattices: Morphological changes. *Cell Motil Cytoskel*, 40(1):13–21, 1998.

7. Paul Fishwick, Stephan Diehl, Jane Prophet, and Jonas Lwgren. Perspectives on aesthetic computing. *Leonardo: The Journal of the International Society for The Arts, Sciences and Technology*, In Press.

8. et al. Gonzalez, P. Cellulat: an agent-based intracellular signalling model. *BioSystems*, pages 171–185, 2003.

9. Jan Hilgevoord and Jos Uffink. The uncertainty principle. In Edward N. Zalta, editor, *The Stanford Encyclopedia of Philosophy*. Winter 2001.

10. M. Kirkland. A phase space model of hemopoiesis and the concept of stem cell renewal. *Exp Hematol*, 32:511–519, 2004.

11. K. Knoepsel. Information visualization. In *Approaching Cyberculture: Humanists as Actors in the Development of Technology confere nce*, Blekinge Institute of Technology and Georgia Institute of Technology, Karlskrona, Sweden, 2003.

12. J. Kosuth. *Art After Philosophy and After: Collected Writings 1966-1990, (G, Gabriele editor)*. MIT Press, 1991.

13. D. S. Krause, N. D. Theise, M. I. Collector, O. Henegariu, S. Hwang, R. Gardner, S. Neutzel, and S. J. Sharkis. Multi-organ, multi-lineage engraftment by a single bone marrow-derived stem cell. *Cell*, 105:369–77, 2001.

14. Sol Lewitt. *Paragraphs on Conceptual Art*. Artforum, Summer Issue, 1967.

15. M. Loeffler and I. Roeder. Tissue stem cells: definition, plasticity, heterogeneity, self-organization and models – a conceptual approach. *Cells Tissues Organs*, 171:8–26, 2002.

16. M. Luck and M. d'Inverno. A conceptual framework for agent definition and development. *The Computer Journal*, 44(1):1–20, 2001.

17. Michael Luck, Ronald Ashri, and Mark d'Inverno. *Agent-Based Software Development*. Artech House, 2004.

18. Marcus Novak. Alienspace. *http://www.mat.ucsb.edu/ marcos/CentrifugeSite/MainFrameSet.html*, 2004.

19. J. Prophet. Sublime ecologies and artistic endeavours: artificial life, interactivity and the internet. *Leonardo: The Journal of the International Society for the Arts, Sciences & Technology*, 29(5), 1996.

20. J. Prophet and M. d'Inverno. Transdisciplinary research in cell. In Paul Fishwick, editor, *Aesthetic Computing*. MIT Press, to appear, 2004.

21. J. Prophet and P. Ride. Wonderful: Visions of the near future book. In M. Fusco, editor, *Active Daydreaming: the nature of collaboration*, pages 66–71. BKD Special Projects, 2001.

22. Jane Prophet. Staining space image 1. *http://www.janeprophet.co.uk/staining03.html and http://www.janeprophet.co.uk/CellApplet02/CellApplet100.html*, 2004.

23. Jane Prophet. Staining space image 2. *http://www.janeprophet.co.uk/cellp01.html*, 2004.

24. Jane Prophet. Staining space image 3. *http://www.janeprophet.co.uk/cellp03.html*, 2004.

25. Jane Prophet. Staining space image 4. *http://www.janeprophet.co.uk/staining01.html*, 2004.

26. I. Roeder. Dynamical modelling of hematopoietic stem cell organisation. *Ph.D. Dissertation Leipzig University*, 2003.
27. I. Roeder and M. Loeffler. A novel dynamic model of hematopoietic stem cell organization based on the concept of within-tissue plasticity. *Exerimental Hematology*, 30:853–861, 2002.
28. J. M. Spivey. *The Z Notation: A Reference Manual*. Prentice-Hall, 1992.
29. N. D. Theise. New principles of cell plasticity. *C R Biologies*, 325:1039–1043, 2003.
30. N. D. Theise and M. d'Inverno. Understanding cell lineages as complex adaptive systems. *Blood, Cells, Molecules and Diseases*, 32:17–20, 2003.
31. N. D. Theise and D. S. Krause. Toward a new paradigm of cell plasticity. *Leukemia*, 16:542–548, 2002.
32. S. Viswanathan and P. Zandstra. Toward predictive models of stem cell fate. *Cryotechnology Review*, 41(2):1–31, 2004.

Statistical Model Selection Methods Applied to Biological Networks

Michael P.H. Stumpf[1,*], Piers J. Ingram[1], Ian Nouvel[1], and Carsten Wiuf[2]

[1] Centre for Bioinformatics, Department of Biological Sciences, Wolfson Building,
Imperial College London, London SW7 2AZ, UK
[2] Bioinformatics Research Center, University of Aarhus, 8000 Aarhus C, Denmark
m.stumpf@imperial.ac.uk

Abstract. Many biological networks have been labelled scale-free as their degree distribution can be approximately described by a powerlaw distribution. While the degree distribution does not summarize all aspects of a network it has often been suggested that its functional form contains important clues as to underlying evolutionary processes that have shaped the network. Generally determining the appropriate functional form for the degree distribution has been fitted in an ad-hoc fashion.

Here we apply formal statistical model selection methods to determine which functional form best describes degree distributions of protein interaction and metabolic networks. We interpret the degree distribution as belonging to a class of probability models and determine which of these models provides the best description for the empirical data using maximum likelihood inference, composite likelihood methods, the Akaike information criterion and goodness-of-fit tests. The whole data is used in order to determine the parameter that best explains the data under a given model (*e.g.* scale-free or random graph). As we will show, present protein interaction and metabolic network data from different organisms suggests that simple scale-free models do not provide an adequate description of real network data.

1 Introduction

Network structures which connect interacting particles such as proteins have long been recognised to be linked to the underlying dynamic or evolutionary processes[13, 3]. In particular the technological advances seen in molecular biology and genetics increasingly provide us with vast amounts of data about genomic, proteomic and metabolomic network structures [15, 22, 19]. Understanding the way in which the different constituents of such networks, — genes and their protein products in the case of genome regulatory networks, enzymes and metabolites in the case of metabolic networks (MN), and proteins in the case of protein interaction networks (PIN) — interact can yield important insights into basic biological mechanisms [16, 24, 1]. For example the extent of

---

[*] Corresponding author.

C. Priami et al. (Eds.): Trans. on Comput. Syst. Biol. III, LNBI 3737, pp. 65–77, 2005.

phenotypic plasticity allowed for by a network, or levels of similarity between PINs in different organisms presumably depend on topological (in a loose sense of the word) properties of networks.

Our analysis here focuses on the degree distribution of a network, *i.e.* the probability of a node to have k connections to other nodes in the network. While it is well known that this does not offer an exhaustive description of network data, it has nevertheless remained an important characteristic/summary statistic of network data. Here we use $\Pr(k)$ to denote a theoretical model for the degree distribution, or $\Pr(k;\theta)$ if the model depends on an (unknown) parameter θ (potentially vector-valued), and $\hat{\Pr}(k)$ to denote the empirical degree distribution.

Many studies of biological network data have suggested that the underlying networks show scale-free behaviour [8] and that their degree distributions follow a power-law, *i.e.*

$$\Pr(k;\gamma) = k^{-\gamma}/\zeta(\gamma) \tag{1}$$

where $\zeta(x)$ is Riemann's zeta-functions which is defined for $x > 1$ and diverges as $x \to 1 \downarrow$; for finite networks, however, it is not necessary that the value of γ is restricted to values greater than 1.

These powerlaws are in marked contrast to the degree distribution of the Erdös-Rényi random graphs [7] which is Poisson, $\Pr(k;\lambda) = e^{-\lambda}\lambda^k/k!$. The study of random graphs is a rich field of research and many important properties can be evaluated analytically. Such Poisson random networks (PRN) are characterized by most nodes having comparable degree; the vast majority of nodes will have a connectivity close to the average connectivity.

The term "scale-free" means that the ratio $\Pr(\alpha k)/\Pr(k)$ depends on α alone but not on the connectivity k. The attraction of scale-free models stem from the fact that some simple and intuitive statistical models of network evolution cause powerlaw degree distribution. Scale-free networks are not, however, the only type of network that produces fat-tailed degree distributions.

Here we will be concerned with developing a statistically sound approach for inferring the functional form for the degree distribution of a real network. We will show that relatively basic statistical concepts, like maximum likelihood estimation and model selection can be straightforwardly applied to PIN and MN data. In particular we will demonstrate how we can determine which probability models best describe the degree distribution of a network. We then apply this approach in the analysis of real PIN data from five model organisms and MN data. In each case we can show that the explanatory power of a standard scale-free network is vastly inferior compared to models that take the finite size of the system into account.

2   Statistical Tools for the Analysis of Network Data

Here we are only concerned with methods aimed at studying the degree distribution of a network. In particular we want to quantify the extent to which a given functional form can describe the degree distribution of a real network. Given a probability model (*e.g.* power-law distribution or Poisson distribution) we want to determine the parameters which describe the degree distribution best; after that we want to be able to distinguish which model from a set of trial model provides the best description. Here we briefly introduce the basic statistical concepts employed later. These can be found in much greater detail in most modern statistics texts such as [12]. Tools for the analysis of other aspects of network data, *e.g.* cluster coefficients, path length or spectral properties of the adjacency matrix will also need to be developed in order to understand topological and functional properties of networks.

There is a well established statistical literature that allows us to assess to what extent data (*e.g.* the degree distribution of a network) is described by a specific probability model (*e.g.* Poisson, exponential or powerlaw distributions). Thus far, determining the best model appears to have been done largely by eye [13] and it is interesting to apply a more rigorous approach, although in some published cases maximum likelihood estimates were used to determine the value of γ for the scale-free distribution.

2.1 Maximum Likelihood Inference

Since we only specify the marginal probability distribution, *i.e.* the degree distribution, we take a composite likelihood approach to inference, and treat the degrees of nodes as independent observations. This is only correct in the limit of an infinite sized network and finite sized sample ($n << N$, where N denotes network size and n the sample size). Composite likelihood methods are becoming increasingly popular in cases for which the full likelihood is difficult to specify and/or the full likelihood is intractable to calculate numerically. In our case

Table 1. Network models and their degree distributions, $\Pr(k; \theta)$. Wherever it appears, C denotes the normalizing constant such that $\sum_k \Pr(k; \theta) = 1$.

Network type	Degree distribution $\Pr(k; \theta)$		Model
Poisson	$\exp(-\lambda)\frac{\lambda^k}{k!}$	for all $k \geq 0$	**M1**
Exponential	$C \exp(-k/\bar{k})$	for all $k \geq 0$	**M2**
Gamma	$\frac{k^{\gamma-1}e^{-k}}{\Gamma(\gamma)}$	for all $k \geq 0$	**M3**
Scale-free	0	for $k = 0$	**M4**
	$k^{-\gamma}/\zeta(\gamma)$	for $k > 0$	
Truncated scale free network	0	for $k < L$ and $k > M$	**M4a**
	$k^{-\gamma}/\sum_{i=L}^{M} k^{-\gamma}$	for $L \leq k \leq M$	
Scale-free network with exponential cut-off	0	for $k < k_0$ and $k > k_{\text{cut}}$	**M4b**
	$(k + k_0)^{-\gamma}\exp(-k/k_{\text{cut}})$	for $k_0 \leq k \leq k_{\text{cut}}$	
Lognormal	$C\frac{e^{-\ln((k-\theta)/m)^2/(2\sigma^2)}}{(k-\theta)\sigma\sqrt{2\pi}}$	for all $k \geq 0$	**M5**
Stretched exponential	0	for $k < 0$	**M6**
	$C \exp(-\alpha k/\bar{k})k^{-\gamma}$	for $k > 0$	

the full likelihood is difficult to specify. Reference [11] provides an overview of composite likelihood methods.

For a given functional form or model $\Pr(k;\theta)$ of the degree distribution we can use maximum likelihood estimation applied to the composite likelihood in order to estimate the parameter which best characterizes the distribution of the data. The composite likelihood of the model given the observed data $K = \{k_1, k_2, \ldots, k_n\}$ is defined by

$$L(\theta) = \prod_{i=1}^{n} \Pr(k_i; \theta), \tag{2}$$

and taking the logarithm yields the log-likelihood

$$\mathrm{lk}(M) = \mathrm{lk}(\theta) = \sum_{i=1}^{n} \log(\Pr(k_i; \theta)). \tag{3}$$

The maximum likelihood estimate (MLE), $\hat{\theta}$, of θ is the value of θ for which Eqns. (2) and (3) become maximal. For this value the observed data is more probable to occur than for any other parameters.

Here the maximum likelihood framework is applied to the whole of the data. This means that in fitting a curve —such as a powerlaw $k^{-\hat{\gamma}}/\zeta(\hat{\gamma})$, where $\hat{\gamma}$ denotes the MLE of the exponent γ— data for all k is considered. If a powerlaw-dependence where to exist only over a limited range of connectivities then the global MLE curve may differ from such a localized power-law (or equivalently any other distribution).

2.2   Model Selection and Akaike Weights

We are interested in determining which model describes the data best. For non-nested models (as are considered here, *e.g.* scale-free versus Poisson) we cannot use the standard likelihood ratio test but have to employ a different information criterion to distinguish between models: here we use the Akaike-information criterion (AIC) to choose between different models [2, 10]. The AIC for a model $\Pr(k;\theta)$ is defined by

$$\mathrm{AIC} = 2(-\mathrm{lk}(\hat{\theta}) + d), \tag{4}$$

where $\hat{\theta}$ is the maximum liklihood estimate of θ and d is the number of parameters required to define the model, *i.e.* the dimension of θ. Note that the model is penalized by d. The model with the minimum AIC is chosen as the best model and the AIC therefore formally biases against overly complicated models. A more complicated model is only accepted as better if it contains more information about the data than a simpler model. (It is possible to formally derive the AIC from Kohn-Sham information theory.) Other information criteria exist, *e.g.* the Bayesian information criterion (BIC) offers a further method for penalizing more complex models (*i.e.* those with more parameters) unless they have significantly higher explanatory power (see [10] for details about the AIC and model selection in statistical inference). In order to compare different models we define the relative differences

$$\Delta_j^{\text{AIC}} = \text{AIC}_j - \min_j(\text{AIC}), \tag{5}$$

where j refers to the jth model, $j = 1, 2, \ldots, J$, and $\min_j$ is minimum over all j. This in turn allows us to calculate the relative likelihoods (adjusted for the dimension of θ) of the different models, given by

$$\exp(-\Delta_j^{\text{AIC}}/2). \tag{6}$$

Normalizing these relative likelihoods yields the so-called Akaike weights w_j,

$$w_j = \frac{\exp(-\Delta_j^{\text{AIC}}/2)}{\sum_{j=1}^J \exp(-\Delta_j^{\text{AIC}}/2)}. \tag{7}$$

The Akaike weight w_j can be interpreted as the probability that model j (out of the J alternative models) is the best model given the observed data and the range of models to choose from. The relative support for one model over another is thus given by the ratio of their respective Akaike weights. If a new model is added to the J existing models then the analysis has to be repeated. The Akaike weight formalism is very flexible and has been applied in a range of context including the assessment of confidence in phylogenetic inference [20]. In the next section we will apply this formalism to PIN data from five species and estimate the level of support for each of the models in table 1.

2.3   Goodness-of-Fit

In addition to the AIC or similar information criteria we can also assess a model's performance at describing the degree distribution using a range of other statistical measures. The Kolmogorov-Smirnoff (KS)[12] and Anderson-Darling (AD) [4, 5] goodness-of-fit statistics allow us to quantify the extent to which a theoretical or estimated model of the degree distribution describes the observed data. The former is a common and easily implemented statistic, but the latter puts more weight on the tails of distributions and also allows for a secular dependence of the variance of the observed data on the argument (here the connectivity k). They KS statistic is defined as

$$D = \max |\hat{P}(k) - P(k)|, \tag{8}$$

where $\hat{P}(k)$ and $P(k)$ are the empirical and theoretical cumulative distribution functions, respectively, for a node's degree i.e. $P(k) = \sum_{i=1}^k \Pr(i)$ and $\hat{P}(k) = \sum_{i=1}^k \hat{\Pr}(i)$. If $P(k)$ depends on θ, $P(k)$ is substituted by $P(k; \hat{\theta}) = \sum_{i=1}^k \Pr(i; \hat{\theta})$, the estimated cumulative distribution. This statistic is most sensitive to differences between the theoretical (or estimated) and observed distributions around the median of the data, i.e. the point where $P(k) \approx 0.5$. Given that we will also be considering a number of fat-tailed distributions this is somewhat unsatisfactory and we will therefore also use the AD statistic (Anderson and Darling discussed a number of statistics [4, 5]) which is defined as

$$D^* = \max \frac{|\hat{P}(k) - P(k)|}{\sqrt{P(k)(1 - P(k))}} \tag{9}$$

(again $P(k)$ might be substituted for $P(k; \hat{\theta})$).

We can use these statistics for two purposes: first, we can use them to compare different trial distributions as the "best" distribution should have the smallest value of D and D^*, respectively. Second, we can use these statistics to determine if the empirical degree distribution is consistent with a given theoretical (or estimated) distribution.

To evaluate the fit of a model, p-values can be calculated for the observed values of D and D^* using a parametric boot-strap procedure using the estimated degree distribution: for a network with N nodes we repeatedly sample N values at random from the maximum likelihood model, $\Pr(k, \hat{\theta})$ and calculate D^* and D^*, respectively, for each replicate. From L bootstrap-replicates we obtain the Null distribution of D and D^* under the estimated degree distribution. For sufficiently large L we can thus determine approximate p-values which allow us to test if the empirical degree distribution is commensurate with the estimated degree distribution.

3   Statistical Analysis of Biological Networks

Here we apply the analysis of the preceeding sections to the study of PINs and metabolic networks. It is easy to find a straight-line fit to some degree interval for all of the datasets considered here. For a powerlaw to be meaningful it has to extend over at least two or three decades and with a maximum degree of $k_{\max} \lesssim 300$ this will be unachievable for the present data sets. We therefore use all the data and fit the model which yields the best overall a description of the degree distribution.

3.1   Analysis of PIN Data

In table 2 we show the maximum composite likelihoods for the degree distributions calculated from PIN data collected in five model organisms [23] (the protein interaction data was taken from the DIP data-base; http://dip.doe-mbi.ucla.edu). We find that the standard scale-free model (or its finite size versions) never provides the best fit to the data; in three networks (*C.elegans, S.cerevisiae* and *E.coli*) the lognormal distribution (M5) explains the data best. In the remaining two organisms the stretched exponential model provides the best fit to the data. The bold likelihoods correspond to the highest Akaike weights. Apart from the case of *H.pylori* (where $\max(w_j) = w_5 \approx 0.95$ for M5 and $w_6 \approx 0.05$) the value of the maximum Akaike weight is always > 0.9999. For *C elegans*, however, the scale-free model and its finite size versions are better than the lognormal model, M5.

Table 2. Log-likelihoods for the degree distributions from table 1 in five model organisms. M1, M2, M3 and M4 have one free parameter, M4a, M4b and M5 have two free parameters, while M6 has three free parameters. The differences in the log-likelihoods are, however, so pronounced that the different numbers of parameters do not noticeably influence the AIC.

Organism	M1	M2	M3	M4	M4a	M4b	M5	M6
D.melanogaster	-38273	-20224	-29965	-18517	-18257	-18126	-17835	**-17820**
C.elegans	-9017	-5975	-6071	-4267	-4258	-4252	-4328	**-4248**
S.cerevisiae	-24978	-14042	-20342	-13454	-13281	-13176	**-12713**	-12759
H. pylori	-2552	-1776	-2052	-1595	-1559	-1546	**-1527**	-1529
E.coli	-834	-884	-698	-799	-789	-779	**-659**	-701

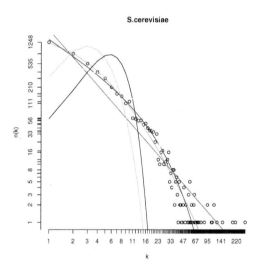

Fig. 1. Yeast protein interaction data (o) and best-fit probability distributions: Poisson (—), Exponential ( ), Gamma (—), Power-law (—), Lognormal (—), Stretched exponential (—). The parameters of the distributions shown in this figure are the maximum likelihood estimates based on the real observed data.

For the yeast PIN the best fit curves (obtained from the MLEs of the parameters of models M1-M6) are shown in figure 1, together with the real data. Visually, log-normal (green) and stretched exponential (blue) appear to describe the date almost equally well. Closer inspection, guided by the Akaike weights, however, shows that the fit of the lognormal to the data is in fact markedly better than the fit of the stretched exponential. But the failure of quickly decaying distributions such as the Poisson distribution, characteristic for classical random graphs [7] to capture the behaviour of the PIN degree distribution is obvious.

Interestingly, common heuristic finite size corrections to the standard scale-free model improve the fit to the data (measured by the AIC). But compared to the lognormal and stretched exponential models they still fall short in describing the PIN data in the five organisms.Figure 2 shows only the three curves with

Table 3. The KS and AD statistics D and D^* for the scale-free, lognormal and stretched exponential model. The ordering of these models is in agreement with the AIC, but KS and AD statistics capture different aspects of the degree distribution. We see that the likelihood treatment, which takes in all the data, agrees better with the KS statistic. In the tails (especially for large connectivities k) the maximum likelihood fit sometimes —especially in the case of *E.coli*— can provide a very bad description of the observed data.

Species	M4		M5		M6	
	D	D^*	D	D^*	D	D^*
D.melanogaster	0.13	0.26	0.01	0.06	0.02	0.06
C.elegans	0.03	0.09	0.10	0.20	0.02	0.08
S.cerevisiae	0.17	0.33	0.01	0.04	0.03	5.99
H. pylori	0.13	0.26	0.01	0.05	0.02	0.12
E.coli	0.28	0.56	0.04	56.09	0.12	6072

the highest values of ω_j, which apart from *E.coli* are the log-normal, stretched exponential and power-law distributions; for *E.coli*, however, the Gamma distribution replaces the power-law distribution. These figures show that, apart from *C.elegans* the shape of the whole degree distribution is not power-law like, or scale-free like, in a strict sense. Again we find that log-normal and stretched exponential distributions are hard to distinguish based on visual assessment alone. Figures 1 and 2, together with the results of table 2, reinforce the well known point that it is hard to choose the best fitting function based on visual inspection. It is perhaps worth noting, that the PIN data is more complete for *S.cerevisiae* and *D.melanogaster* than for the other organisms.

The standard scale-free model is superior to the log-normal only for *C.elegans*. The order of models (measured by decreasing Akaike weights) is M6, M5, M4, M2, M3, M1 for *D.melanogaster*, M6, M4, M5, M2, M3, M1 for *C.elegans*, M5, M6, M4, M2, M3, M1 for *S.cerevisiae* and *H.pylori*, and M5, M3, M6, M4, M2, M1 for *E.coli*. Thus in the light of present data the PIN degree distribution of *E.coli* lends more support to a Gamma distribution than to a scale-free (or even stretched scale-free) model. There is of course, no mechanistic reason why the gamma distribution should be biologically plausible but this point demonstrates that present PIN data is more complicated than predicted by simple models. Therefore statistical model selection is needed to determine the extent to which simple models really provide insights into the intricate architecture of PINs. For

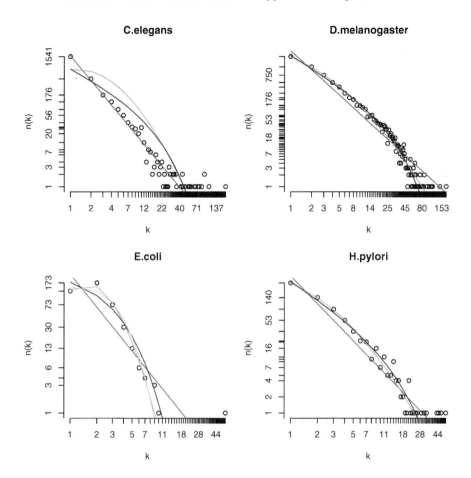

Fig. 2. Degree distributions of the protein interaction networks (o) of *C.elegans*, *D.melanogaster*, *E.coli* and *H.pylori*. The power-law (—), log-normal (—) and stretched exponential (—) models are shown for all figures; for *E.coli* the gamma distribution (—), which performs better (measured by the Akaike weights) than either scale-free and the stretched exponential distributions.

completeness we note that model selection based on BIC results in the same ordering of models as the AIC shown here.

In table 3 we give the values of D and D^* for the empirical degree distributions. The estimated cumulative distribution $P(k; \hat{\theta})$ was obtained from the maximum likelihood fits of the respective models. The results in table 3 show that the maximum likelihood framework (or the respective models) sometimes cannot adequately describe the tails of the distribution —at low and high values of the connectivity— in some cases, where $D^* >> D$. The order of the different models suggested by D for the three models generally agrees with the ordering obtained from the AIC.

3.2   Analysis of MN Data

Metabolic networks aim to describe the biochemical machinery underlying cellular processes. Here we have used data from the KEGG database (www.genome. jp/kegg) with additional information from the BRENDA database (www. brenda.uni-koeln.de). The nodes are the enzymes and an edge is added between two enzymes if one uses the products of the other as educts.

Table 4. Log-likelihoods for the degree distributions from table 1 applied to metabolic networks. Human and yeast data were extracted from the global database.

Data	M1	M2	M3	M4	M5	M6
KEGG	-8276	-5247	-7676	-5946	-5054	**-4178**
H.sapiens	-1619	-1390	-1565	-1525	-1312	**-1308**
S.cerevisiae	-2335	-1485	-2185	-1621	-1436	**-1427**

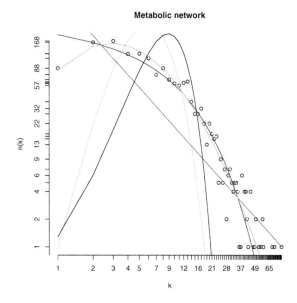

Fig. 3. Degree distributions of the metabolic network data (o) and best-fit probability distributions: Poisson (—), Exponential ( ), Gamma (—), Power-law (—), Lognormal (—), Stretched exponential (—). The parameters of the distributions shown in this figure are the maximum likelihood estimates based on the real observed data. Ignorig low degrees ($k = 1, 2$) when fitting the scale-free model increases the exponent (*i.e.* it falls off more steeply) but compared to the other models no increased performance is observed.

From figure 3.2 and table 4 it is apparent that the maximum likelihood scale-free model obtained from the whole network data does not provide an adequate description of the MN data. This should, however, not be too surprising as the network is only relatively small with maximum degree $k_{max} = 83$. The degree distribution appears to decay in an essentially exponential fashion but the stretched exponential has the required extra flexibility to describe the whole of the data better than the other models. Measured by goodness of fit statistics D and D^*, however, the log-normal model ($D = 0.02$ and $D^* = 0.06$) performs rather better than the stretched exponential ($D = 0.07$ and $D = 0.22$); the scale-free model again performs very badly ($D = 1.0$ and $D^* = 34.2$) for the metabolic network data.

4   Conclusions

We have shown that it is possible to use standard statistical methods in order to determine which probability model describes the degree distribution best. We find that the common practice of fitting a pure power-law to such experimental network data[13, 3] may obscure information contained in the degree distribution of biological networks. This is often done by identifying a range of connectivities from the log-log plots of the degree distribution which can then be fitted by a straight line. Not only is this wasteful in the sense that not all of the data is used but it may obfuscate real, especially finite-size, trends. The same will very likely hold true for other biological networks, too [17]. The approach used here, on the other hand, (i) uses all the data, and (ii) can be extended to assessing levels of confidence through combining a bootstrap procedure with the Akaike weights. What we have shown here, in summary, is that statistical methods can be usefully applied to protein interaction and metabolic network data.

Even though the degree distribution does not capture all (or even most) of the characteristics of real biological networks there is reason to reevaluate previous studies. We find that formally real biological networks provide very little support for the common notion that biological networks are scale-free. Other fat-tailed probability distributions provide qualitatively and quantitatively better descriptions of the degree distributions of biological networks. For protein interaction networks we found that the log-normal and the stretched exponential offer superior descriptions of the degree distribution than the powerlaw or its finite size versions. For metabolic versions our results confirms this. Even the exponential model outperformed the scale-free model in describing the empirical degree distribution (we note that randomly growing networks are characterized by an exponentially decreasing degree distribution). The best models are all fat-tailed —like the scale-free models— but are not formally scale-free. Unfortunately, there is as yet no known physical model for network growth processes that would give rise to log-normal or stretched exponential degree distributions.

There is thus a need to develop and study theoretical models of network growth that are better able to describe the structure of existing networks. This probably needs to be done in light of at least three constraints: (i) real networks are finite sized and thus, in the terms of statistical physics, mesoscopic systems; (ii) present network data are really only samples from much larger networks as not all proteins are included in present experimental setup (in our case *S.cerevisiae* has the highest fraction, 4773 out of approximately 5500-6000 proteins); the sampling properties have recently been studied and it was found that generally the degree distribution of a subnet will differ from that of the whole network. This is particularly true for scale-free networks. (iii) biological networks are under a number of functional and evolutionary constraints and proteins are more likely to interact with proteins in the same cellular compartment or those involved in the same biological process. This modularity —and the information already availabe *e.g.* in gene ontologies— needs to be considered. Finally there is an additional caveat: biological networks are not static but likely to change during development. More dynamic structures may be required to deal with this type of problem.

Quite generally we believe that we are now at a stage where simple models do not necessarily describe the data collected from complex processes to the extent that we would like them to. But as Burda, Diaz-Correia and Krzywicki point out [9], even if a mechanistic model is not correct in detail, a corresponding statistical ensemble may nevertheless offer important insights. We believe that the statistical models employed here will also be useful in helping to identify more realistic ensembles.

The maximum likelihood, goodness of fit and other tools and methods for the analysis of network data are implemented in the NetZ R-package which is available from the corresponding author on request.

Acknowledgements. We thank the Wellcome Trust for a research fellowship (MPHS) and a research studentship (PJI). CW is supported by the Danish Cancer Society. Financial support from the Royal Society and the Carlsberg Foundation (to MPHS and CW) is also gratefully acknowledged. We have furthermore benefitted from discussions with Eric de Silva, Bob May and Mike Sternberg.

References

[1] I. Agrafioti, J. Swire, J. Abbott, D. Huntley, S. Butcher and M.P.H. Stumpf. Comparative analysis of the *Saccharomyces cerevisiae* and *Caenorhabditis elegans* protein interaction networks. *BMC Evolutionary Biology*, 5:23, 2005.

[2] H. Akaike. Information measures and model selection. In *Proceedings of the 44th Session of the International Statistical Institute*, pages 277–291, 1983.

[3] R. Albert and A. Barabási. Statistical mechanics of complex networks. *Rev. Mod. Phys.*, 74:47, 2002.

[4] T.W. Anderson and D.A. Darling Asymptotic theory of certain goodness-of-fit criteria based on stochastc processes. *Ann.Math.Stat.*, 23:193, 1952.

[5] T.W. Anderson and D.A. Darling A test of goodness of fit *J.Am.Stat.Assoc.*, 49:765, 1954.

[6] A. Barabási and R. Albert. Emergence of scaling in random networks. *Science*, 286:509, 1999.

[7] B. Bollobás. *Random Graphs*. Academic Press, London, 1998.

[8] B. Bollobás and O. Riordan. Mathematical results on scale-free graphs. In S. Bornholdt and H. Schuster, editors, *Handbook of Graphs and Networks*, pages 1–34. Wiley-VCH, 2003.

[9] Z. Burda and J. Diaz-Correia and A. Krzywicki. Statistical ensemble of scale-free random Graphs. *Phys. Rev. E*, 64,:046118,2001.

[10] K. Burnham and D. Anderson. *Model selection and multimodel inference: A practical information-theoretic approach.* Springer, 2002.

[11] D. R. Cox and Reid. A note on pseudolikelihood constructed from marginal densities. *Biometrika*, 91:729, 2004.

[12] A. Davison. *Statistical models.* Cambridge University Press, Cambridge, 2003.

[13] S. Dorogovtsev and J. Mendes. *Evolution of Networks.* Oxford University Press, Oxford, 2003.

[14] S. Dorogovtsev, J. Mendes, and A. Samukhin. Multifractal properties of growing networks. *Europhys. Lett.*, 57:334;cond–mat/0106142, 2002.

[15] T. Ito, K. Tashiro, S. Muta, R. Ozawa, T. Chiba, M. Nishizawa, K. Yamamoto, S. Kuhara, and Y. Sakaki. Towards a protein-protein interaction map of the budding yeast: A comprehensive system to examine two-hybrid interactions in all possible combinations between the yeast proteins. *PNAS*, 97:1143, 2000.

[16] S. Maslov and K. Sneppen. Specificity and stability in topology or protein networks. *Science*, 296:910, 2002.

[17] R. May and A. Lloyd. Infection dynamics on scale-free networks. *Phys. Rev. E*, 64:066112, 2001.

[18] M. Newman, S. Strogatz, and D. Watts. Random graphs with arbitrary degree distribution and their application. *Phys. Rev. E*, 64:026118;cond–mat/0007235, 2001.

[19] H. Qin, H. Lu, W. Wu, and W. Li. Evolution of the yeast interaction network. *PNAS*, 100:12820, 2003.

[20] K. Strimmer and A. Rambaut. Inferring confidence sets of possibly misspecified gene trees. *Proc. Roy. Soc. Lond. B*, 269:127, 2002.

[21] M.P.H. Stumpf, C. Wiuf and R.M. May Subnets of scale-free networks are not scale-free: the sampling properties of random networks. *PNAS*, 103:4221, 2005.

[22] A. Wagner. The yeast protein interaction network evolves rapidly and contains few redundant duplicate genes. *Mol. Biol. Evol.*, 18:1283, 2001.

[23] I. Xenarios, D. Rice, L. Salwinski, M. Baron, E. Marcotte, and D. Eisenberg. Dip: the database of interacting proteins. *Nucl. Acid. Res.*, 28:289, 2000.

[24] S. Yook, Z. Oltvai, and A. Barabási. Functional and topological characterization of protein interaction networks. *Proteomics*, 4:928, 2004.

Using Secondary Structure Information to Perform Multiple Alignment

Giuliano Armano[1], Luciano Milanesi[2], and Alessandro Orro[1]

[1] University of Cagliari, Piazza d'Armi, I-09123, Cagliari, Italy
{armano, orro}@diee.unica.it
http://iasc.diee.unica.it
[2] ITB-CNR, Via Fratelli Cervi, 93, 20090 Segrate Milano, Italy
luciano.milanesi@itb.cnr.it
http://www.itb.cnr.it

Abstract. In this paper an approach devised to perform multiple alignment is described, able to exploit any available secondary structure information. In particular, given the sequences to be aligned, their secondary structure (either available or predicted) is used to perform an initial alignment –to be refined by means of locally-scoped operators entrusted with "rearranging" the primary level. Aimed at evaluating both the performance of the technique and the impact of "true" secondary structure information on the quality of alignments, a suitable algorithm has been implemented and assessed on relevant test cases. Experimental results point out that the proposed solution is particularly effective when used to align low similarity protein sequences.

1 Introduction

Alignment is the most acknowledged method for comparing two or more proteins, highlighting their relations throughout insertion and deletion of aminoacids. In this way, both local and global homologies can be detected. In fact, structurally similar sequences have high likelihood of being similar also from a functional point of view. In addition, the similarity among sequences pertaining to different organisms can help in highlighting evolutionary relationships among the involved species.

The problem of calculating pairwise sequence alignments can be solved using dynamic programming [15], a widely acknowledged technique that guarantees the optimality of the result. On the other hand, the high computational complexity associated with multiple alignment does not guarantee that optimal results can always be found, the problem being NP-complete. In fact, also considering experiments reported in [3], it may be asserted that dynamic programming on multiple sequences is effective only for a small number of sequences. In this case, suitable techniques aimed at identifying alignments that are likely to be near to the optimal one should be adopted. Roughly speaking, multiple alignment techniques can be classified in two main categories: progressive and iterative –see also [28] and [19][1] for further details.

---

[1] Notredame considers also the so-called "exact" techniques. Although conceptually relevant, this further category has been disregarded for the sake of simplicity.

C. Priami et al. (Eds.): Trans. on Comput. Syst. Biol. III, LNBI 3737, pp. 78–88, 2005.
© Springer-Verlag Berlin Heidelberg 2005

Progressive techniques [6] [7] –often related with pairwise alignment performed through dynamic programming– build the alignment incrementally, so that no change is performed on sequences previously taken into account. The CLUSTAL algorithm [27] is a typical example of these techniques. It builds a protein tree according to a distance measure evaluated between sequence pairs. The tree establishes a partial order, to be followed while integrating sequences step-by-step. PIMA [25] uses a similar approach, the most important difference being the fact that sequences are aligned according to the existence of locally-conserved motifs. Two different versions of PIMA have been released, i.e., ML_PIMA and SB_PIMA (standing for Maximum Linkage and Sequential Branching PIMA, respectively), which differ on the way the order of alignment is decided. It is worth pointing out that PIMA may adopt different gap penalties depending on the availability of secondary structure information.

Among other relevant progressive algorithms let us recall MULTAL [26], MULTALIGN [1], and PILEUP [4]. Although very efficient, progressive techniques can hardly escape from local minima, due to the adopted greedy-search strategy. Thus, the most important problem to be solved by progressive techniques is the adoption of the "right" order, to be followed while adding sequences to a partial solution. This problem has also been dealt with by adopting consistency-based objective functions, able to integrate different scoring models [17]. T-Coffee [18] is the most relevant algorithm in this category. At each step, it identifies the sequence to be added using a model based on a library of previously-computed pairwise alignments (both local and global).

To prevent the most recurrent problem of progressive techniques (i.e., that an error performed at a given step cannot be corrected in subsequent steps), iterative techniques allow the adoption of "arrangement" operators that may apply to any sequence in the alignment built so far. Thus, in a sense, iterative techniques are able to enforce a backtracking mechanism, trading the capability of escaping from local minima with a higher computational complexity. In DiAlign [14], multiple alignment is calculated in two steps. A segment being a subsequence with no gaps, first a segment-to-segment comparison is performed on pairs of segments with the same length (called "diagonals"). Then, the alignment is built by joining diagonals according to an iterative procedure. Another relevant algorithm in this category is PRRP [9], which optimises an initial alignment by iteratively calculating weights of all sequence pairs depending on a phylogenetic tree and refining the multiple alignment according to a double-nested strategy. Stochastic iterative techniques exploit non determinism to escape from local minima. In particular, let us recall SAGA [16], which evolves a population of multiple alignments according to a typical evolutionary strategy based on Genetic Algorithms. Arbitrary types of objective functions can be optimized by the algorithm. Among other relevant iterative algorithms let us recall Praline [10], HMMER [13], and HMMT [5].

It is worth noting that most of the methods above do not take into account the context in which comparisons are carried out (a notable exceptions being the position-dependent score model of T-Coffee). A particular kind of contextual information is given by the secondary structure, which labels amino acids depending on their spatial folding. In our opinion, using this information can remarkably help the search for a suboptimal alignment. In fact, a preliminary comparison performed at a structural level can be seen as a particular kind of heuristics –useful to drive the search at the primary level. As an application of this principle, in this paper we exploit the information about secondary structures to obtain an initial alignment, to be used as a starting point for evaluating the final alignment. Assuming that the alignment produced at the secondary level is close to the optimal solution, it is likely that suitable (locally-scoped) arrangements performed at the primary level can lead to satisfactory result with a limited computational cost. An implementation of this technique, called A3 (standing for Abstract Alignment Algorithm), has produced encouraging results on sequences taken from the BAliBASE benchmark database for multiple alignment algorithms [29].

2   Materials and Methods

The informative principle used to attain the objective of reducing the time spent for searching a solution is the adoption of abstraction techniques. Abstraction has already been used in many fields of computer science, i.e., automatic theorem proving [20], [8], planning [12], and learning [24]. Intuitively, abstraction consists of mapping a problem from a concrete to an abstract representation in which some irrelevant details are omitted, so that only the properties considered useful are retained. Working in an abstract space involves various advantages, as finding a solution at an abstract level first, and then refining it at the concrete level, may simplify the overall search.

According to the general principles of abstraction, and given the influence of spatial-folding on the functionality of a protein [11], it appears reasonable to exploit the information about the secondary structure while performing multiple alignments. In particular, sequences are aligned first at the secondary level and then at the primary level, using the upper-level solution as a heuristic devised to limit the complexity of the search. Figure 1 highlights the architecture of A3, consisting of two handlers –one for each level of abstraction. In the current implementation, only a top-down interaction is allowed –i.e., no feedback can be performed by the primary-level handler to its superior. Let us assume that the information about the secondary structure of all sequences be available, a lack of such information being dealt with by predicting it. Figure 2 briefly depicts the algorithm and points to the underlying technique, which interleaves progressive and iterative steps. In fact, the secondary-structure handler enforces an outer cycle that implements a progressive technique, whereas the primary-structure handler enforces an inner cycle that implements an iterative technique.

In principle, the distinction between progressive and iterative techniques is cross-sectional, in the sense that both might be used at either level (i.e., primary or secondary). In practice, also taking into account the complexity of the search, we decided to let the progressive technique be used only by the module that

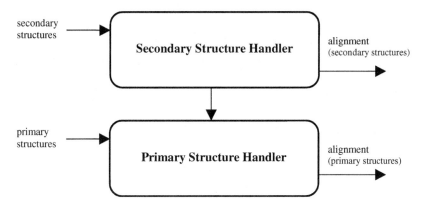

Fig. 1. System Architecture. The secondary level builds a multiple alignment, to be refined by the primary one. The interaction between levels is unidirectional –from the secondary to the primary.

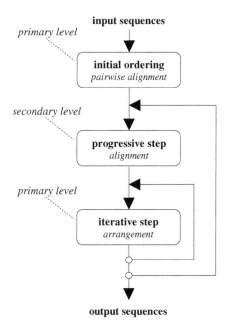

Fig. 2. A3 algorithm. After an initial ordering (at the primary level), sequences are progressively embodied into the solution using dynamic programming (at the secondary level). At each progressive step, the current alignment is iteratively arranged by resorting to locally-scoped operators –entrusted with moving gaps along the sequences (at the primary level).

takes care of the secondary level, whereas the iterative technique is enforced at the primary level.

The progressive technique adds one sequence at a time to the set of output sequences, initially empty. Figure 3 illustrates how sequences are progressively embodied into the current solution, according to the ranking established by an initial ordering –obtained by performing pairwise alignments at the primary level. To this end, at each step, dynamic programming is applied.

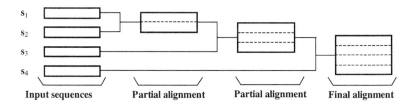

Fig. 3. A graphical representation, showing how sequences are progressively embodied into the solution, is given. The precedence relation being determined in accordance with the result obtained by evaluating their pairwise similarity at the primary level.

The iterative technique takes care of refining the alignment obtained so far, by applying operators devised to increase the value of the selected score function through locally-scoped arrangements. Note that the iterative technique has been adopted to overcome the main drawback of the progressive technique. In fact, if applied as it is, each step of the progressive technique would rule out the possibility of carrying out corrections to the alignment already performed. In this case, any error made in previous steps could possibly be amplified in the next occurring steps. On the contrary, the introduction of the iterative technique greatly helps the search to escape from local maxima of the score function, thanks to its capability of performing local adjustments to the set of output sequences. In particular, an iterative step consists of repeatedly arranging "segments" (here, the term segment denotes a subsequence of amino acids, together with its delimiting gaps). Given a segment, a gap is shifted along it, taking care to preserve the segment's overall length. The score of each resulting alignment is evaluated, and the alignment that obtained the maximum score is selected. A typical action performed while arranging a segment is shown in Figure 4. A more detailed description of the A3 algorithm is reported in Figure 5.

The score model adopted for dealing with secondary structures is based on a substitution matrix derived using superimpositions from protein pairs of similar structure, with low similarity [22]. In accordance with other alignment algorithms, the score model takes into account a cost for gap opening and a cost for each position gap extension; i.e., $g_o + g_e \cdot L$, where L is the gap length.

It is worth pointing out that, sequences have a two-tiered representation, depending on the level used to process them. In particular, at the primary level, a sequence is described as a chain of aminoacids, whereas, at the secondary level, a

DDSFSASD	DDSFSASD	DDSFSASD	DDSFSASD	DDSFSASD	DDSFSASD
D-ASDSA-	DA-SDSA-	DAS-DSA-	DASD-SA-	DASDS-A-	DASDSA--
score=-1 (starting point)	score=-1	score=0	score=0	score=1	score=2
	DDSFSASD	DDSFSASD	DDSFSASD	DDSFSASD	DDSFSASD
	D-ASDS-A	D-ASD-SA	D-AS-DSA	D-A-SDSA	D--ASDSA
	score=-1	score=0	score=0	score=1	score=0

Fig. 4. An example, reporting the behavior of the function that arranges a segment, is given. Gaps are moved along the segment, giving rise to different configurations; the best configuration is selected according to the corresponding score.

Name	A3	
Input	$S = \{s_1, s_2, ..., s_N\}$	Set of sequences to be aligned.
Output	$M = \langle a_1, a_2, ..., a_N \rangle$	List of aligned sequences.
Step 0	$\Sigma = \langle \sigma_1, \sigma_2, ..., \sigma_N \rangle \leftarrow order(S)$	The set of input sequences are ordered, giving rise to the list Σ.
	$M \leftarrow \langle \; \rangle$	The output list of aligned sequences (M) is initially empty.
	$n \leftarrow 1$	Let the index n point to the first sequence in Σ.
Step 1	$M \leftarrow M + \langle \sigma_n \rangle$	Append $\langle \sigma_n \rangle$ to M.
Step 2	$M \leftarrow align2(M)$	_Progressive technique step_ (secondary level): Align the newest element of M with the sequences previously contained in M (by using dynamic programming at the secondary level).
Step 3	$M \leftarrow arrange\_multi(M)$	_Iterative technique step_ (primary level): Apply locally-scoped rearrangement operators to M.
Step 4	$n \leftarrow n + 1$	Let the index n point to the next sequence in Σ.
Step 5	_If $n \leq N$ then goto step 1 else stop_	Repeat until all sequences have been processed.

Name	arrange_multi	
Input	$M = \langle a_1, a_2, ..., a_n \rangle$	List of sequences aligned so far (n<=N).
Output	$M = \langle a'_1, a'_2, ..., a'_n \rangle$	M is modified by the procedure, which uses locally-scoped rearrangement operators.
Step 0	$k \leftarrow |M|$	Let k point to the last sequence in M where $n = |M|$ denotes the cardinality of M.
Step 1	$M \leftarrow arrange\_sequence(k, M)$	All segments of the k-th sequence are arranged according to locally-scoped operators.
Step 2	$M \leftarrow format(M)$	All columns that contain only gaps are removed from M.
Step 3	$k \leftarrow k - 1$	Let k point to the previous sequence in M.
Step 4	_If $k > 0$ then goto step1 else stop_	Repeat until all sequences have been processed.

Fig. 5. Pseudo-code of the A3 algorithm

sequence is described as a chain of secondary-structure labels (i.e., alpha-helices, beta-sheets, and random coils). In so doing the length of the sequence remains unchanged.

At the primary level, for each column, the score function uses a substitution matrix to compare all corresponding residues, and the gap cost is defined as $k_o + k_e \cdot L$. Moreover, small segments are penalized according to the following cost function, devised while tuning the system (where i denotes a generic segment, and l_i its length):

$$30 \cdot \exp\left(-\frac{l_i + 1}{2}\right)$$

The computational complexity of A3 is $O(L^3 N^3)$ where L is the average length of the sequences and N is the number of sequences to be aligned. It is worth pointing out that it is substantially due to the iterative technique, although the time required to perform multiple alignment is usually considerably lower than the one predicted by the above equation, which accounts for the worst case. Furthermore, the equation assumes that the iterative technique is able to enforce globally-scoped arrangement operators, whereas arrangement operators are in fact locally scoped.

3   Results and Discussion

A3 has been tested using alignments taken from BAliBASE, which is organized in five numbered references (denoted as *ref1-ref5*). Each reference contains several groups of sequences to be aligned, together with the corresponding "correct" alignments. Known secondary structures are taken from the database Protein Data Bank (PDB) [2], whereas unknown ones are predicted using SSPRO [21] (alternative secondary-structure predictors could be used, depending on their availability).

The BAliBASE benchmark database has been adopted for testing the algorithm. The quality of alignments has been assessed, for each sequence, by resorting to the *bali_score* program, which measures the "column score" (CS) on "core blocks" selected by the authors of BAliBASE according to their relevance for the functionality of the sequence. Two separate sets of experiments have been performed, aimed at assessing the impact of available secondary structure information on: (i) the accuracy of alignments, and (ii) the capability of handling sequences with low degree of similarity.

In the former set of experiments (see Table 1), only alignments whose amount of available secondary information is greater than a threshold (ranging from 0 to 100%) are considered. Since A3 has been explicitly designed to work with extra information related with secondary structures, results cannot be directly compared with state-of-the-art predictors that perform prediction without resorting to such kind of information. Nevertheless, results obtained running the most widely acknowledged program (i.e., T-Coffee) have been included to let the reader better assess the capabilities of the proposed approach. Table 1 reports also, for each selected threshold (σ), the fraction of alignments (ρ) that satisfy the constraint imposed on the amount of available secondary structure, as well as the difference (Δ) between A3 and T-Coffee.

Table 1. Results obtained by A3 considering only alignments whose available secondary structure information is greater than a threshold (ranging from 0 to 100% of "true" secondary structure information)

$\sigma \geq$	ρ	A3	T-Coffee	Δ
0	100.00%	69.51	70.17	-0.66
10	68.35%	62.92	60.42	2.50
20	55.40%	67.52	64.48	3.04
30	43.88%	68.30	63.97	4.33
40	39.57%	69.04	65.38	3.66
50	37.41%	69.44	65.08	4.36
60	33.09%	68.61	63.80	4.81
70	27.34%	66.50	60.21	6.29
80	21.58%	67.53	59.93	7.60
90	19.42%	67.67	60.26	7.41

The latter set of experiments has been conceived to investigate whether or not information about "true" secondary structures may help in the task of aligning sequences with low similarity. In fact, several experimental results (e.g., [23]) prove that it is a hard task, the majority of multiple alignment programs yielding poor results on alignments with a degree of similarity below 30%. Table 2 reports experimental results obtained by taking into account only alignments whose pairwise similarity is lower than 30%. For the sake of brevity, only "true" secondary structure information in the range 0-30% has been considered. As shown in the table, there is a clear evidence that secondary structure information can help to deal with the "twilight zone" of protein sequence alignments, which indirectly motivates the adoption of the proposed approach under this condition.

Table 2. Results obtained by A3 and T-Coffee on BAliBASE –run on tests that satisfy additional similarity constraints (i.e., pairwise similarity lower than 30%)

$\sigma \geq$	A3	T-Coffee	Δ
0	62.24%	62.09	0.15
5	58.69%	55.76	2.93
10	58.75%	54.88	3.87
15	59.20%	55.52	3.68
20	61.86%	57.84	4.02
25	62.61%	58.20	4.41
30	64.32%	58.32	6.00

4   Conclusion and Future Work

The aim of this work was to assess the impact of secondary structure information in the task of multiple alignment of proteins. To this end, the A3 algorithm,
which exploits secondary structure information to perform multiple alignment,
has been devised and implemented. As expected, experimental results show that
A3 performs better when the amount of secondary structure information is not
negligible (at least 10-20%). Further experiments point out that even a small
amount of extra information allows A3 to perform relatively well below the so-
called "twilight zone" (i.e., about 30% of pairwise similarity between sequences).
As for the future work, we are currently assessing the dependence of arrangement operators on the complexity of the search, so that an automatic trade-off
between complexity and scope of such operators will be enforced according to
the amount of available information at the secondary level. In fact, in the current implementation of the system, the scope of local operators at the primary
level is not related with the amount of available secondary structure. Thus,
an error performed at the secondary level (due to a lack of information about
the actual structure of the sequence) cannot be easily corrected as long as arrangement operators have a limited scope –the less information available, the
more the scope should be augmented. A clustering module (to be used during
preprocessing) is also being implemented, aimed at giving A3 the information
required to cope with groups in *ref3*, characterized by sequences taken from a
limited number of different families. Moreover, we are experimenting different
score models with the aim of improving the performances of A3 with the groups
in *ref5*, characterized by sequences with long internal gaps. Finally, the problem
of how to implement mutual interactions between secondary and primary level
is also being investigated.

Acknowledgments

This work was supported by iINTAS Cyclonet and by MIUR: "Bioinformatics
for Genome and Proteome", "GRID-IT" and Laboratory of Interdisciplinary
Technologies in Bioinformatics (LITBIO) RBLA0332RH , FIRB projects.

References

1. Barton, G.J., Sternberg, M.E.J.: A Strategy for the Rapid Multiple Alignment
   of Protein Sequences. Confidence Levels from Tertiary Structure Comparisons. J.
   Mol. Biol. (1987) 198:327–337
2. Berman, H.M., Westbrook, J., Feng, Z., Gilliland, G., Bhat, T.N., Weissig, H.,
   Shindyalov, I.N., Bourne, P.E.: The Protein Data Bank. Nucleic Acids Research
   (2000) 28:235–242
3. Carrillo, H., Lipman, D.J.: The multiple sequence alignment problem in biology.
   SIAM J. Appl. Math. 48, (1988) 1073–1082
4. Devereux, J., Haeberli, P., Smithies, O.: GCG package. Nucleic Acids Research
   (1984) 12:387–395

5. Eddy, S.R.: Multiple alignment using hidden Markov models. Proc. Int. Conf. Intell. Syst. Mol. Biol. (1995) 3:114–20

6. Feng, D.F., Doolittle, R.F.: Progressive sequence alignment as a prerequisite to correct phylogenetic trees. J. Mol. Evol. (1987) 25:351–360

7. Hogeweg, P., Hesper, B.: The alignment of sets of sequences and the construction of phylogenetic trees, an integrated method. J. Mol. Evol. 20 (1984) 175–186.

8. Giunchiglia, F., Villafiorita, A., Walsh, T.: Theories of Abtraction. AI Communications (1997) 10:167–176

9. Gotoh O.: Significant Improvement in Accuracy of Multiple Protein Sequence Alignments by Iterative Refinement as Assessed by Reference to Structural Alignments. J. Mol. Biol. (1996) 264:823–838

10. Heringa, J.: Two strategies for sequence comparison: profile preprocessed and secondary structure-induced multiple alignment. Computers and Chemistry (1999) 23:341–364

11. Kabsch, W., Sander, C.: Dictionary of protein secondary structure: pattern recognition of hydrogen-bonded and geometrical features. Biopolymers (1983) 22:2577–2637

12. Knoblock, C.A., Tenenberg, J.D., Yang, Q.: Characterizing Abstraction Hierarchies for Planning. Proc. of the Ninth National Conference on Artificial Intelligence (1991) 2:692–697

13. Krogh, A., Brown, M., Mian, I.S., Sjlander, K., Haussler, D.: Hidden Markov Models in Computational Biology: Applications to Protein Modeling. J. Mol. Biol. (1994) 235:1501–1531

14. Morgenstern, B., Dress, A., Werner, T.: Multiple DNA and protein sequence alignment based on segment-to-segment comparison. Proc. Natl. Acad. Sci. USA (1996) 93:12098–12103

15. Needleman, S.B., Wunsch, C.D.: A general method applicable to the search for similarities in the amino acid sequence of two proteins. J. Mol. Biol. (1970) 48:443–453

16. Notredame, C., Higgins, D.G.: SAGA: sequence alignment by genetic algorithm. Nucleic Acids Res. (1996) 24:1515–1524

17. Notredame, C., Holm L., Higgins, D.G.: COFFEE: an objective function for multiple sequence alignments. Bioinformatics (1998) 14:407–422

18. Notredame, C., Higgins, D.G., Heringa J.: T-Coffee: A Novel Method for Fast and Accurate Multiple Sequence Alignment. J. Mol. Biol. (2000) 302:205–217

19. Notredame, C.: Recent Progresses in Multiple Sequence Alignment: a Survey. Pharmaco-genomics. (2002) 3(1):131–144

20. Plaisted, D.: Theorem Proving with Abstraction. Artificial Intelligence (1981) 16(1):47–108

21. Pollastri, G., Przybylski, D., Rost, B., Baldi, P.: Improving the Prediction of Protein Secondary Structure in Three and Eight Classes Using Recurrent Neural Networks and Profiles. Proteins (2002) 47:228–235

22. Prlic, A., Domingues, F.S., Sippl, M.J.: Structure-derived substitution matrices for alignment of distantly related sequences. Protein Eng. (2000) 13:545–550

23. Rost, B.: Twilight zone of protein sequence alignments. Protein Engineering (1999) 12(2):85-94

24. Saitta, L., Zucker, J.D.: Semantic Abstraction for Concept Representation and Learning. Symposium on Abstraction, Reformulation and Approximation (SARA98), Pacific Grove, California (1998) 103–120

25. Smith, R.F., Smith, T.F.: Pattern-Induced Multi-sequence Alignment (PIMA) algorithm employing secondary structure-dependent gap penalties for use in comparative protein modelling. Protein Eng. (1992) 5(1):35–41
26. Taylor, W.R.: A flexible method to align large numbers of biological sequences. J. Mol. Evol. (1988) 28:161–169
27. Thompson, J.D., Higgins, D.G., Gibson, T.J.: CLUSTAL W: improving the sensitivity of progressive multiple sequence alignment through sequence weighting, positionspecific gap penalties, and weight matrix choice. Nucleic Acids Res. (1994) 22:4673–4680
28. Thompson, J.D., Plewniak, F., Poch, O.: A comprehensive comparison of multiple sequence alignment programs. Nucleic Acids Research (1999) 27:2682–2690
29. Thompson, J.D., Plewniak, F., Poch, O.: BAliBASE: a benchmark alignment database for the evaluation of multiple alignment programs. Bioinformatics (1999) 15:87–88

Frequency Concepts and Pattern Detection for the Analysis of Motifs in Networks

Falk Schreiber and Henning Schwöbbermeyer

Leibniz Institute of Plant Genetics and Crop Plant Research,
Corrensstraße 3, D-06466 Gatersleben, Germany
{schreibe, schwoebb}@ipk-gatersleben.de

Abstract. Network motifs, patterns of local interconnections with potential functional properties, are important for the analysis of biological networks. To analyse motifs in networks the first step is to find patterns of interest. This paper presents 1) three different concepts for the determination of pattern frequency and 2) an algorithm to compute these frequencies. The different concepts of pattern frequency depend on the reuse of network elements. The presented algorithm finds all or highly frequent patterns under consideration of these concepts. The utility of this method is demonstrated by applying it to biological data.

1 Introduction

Biological processes form large and complex networks. Network analysis methods help to uncover important properties of these networks and therefore assist scientists in understanding processes in organisms. *Network motifs* are specific patterns of local interconnections with potential functional properties and can be seen as the basic building blocks of complex networks [1]. They are important for the functional analysis of biological networks [2,3]. One example is the *feed-forward loop* motif which has been shown to perform information processing tasks [4].

There are different definitions of a network motif. Some authors use this term to represent a set of related networks [2], for others it is simply a single small network [3]. Often motifs are described as patterns of local interconnections which occur in networks at numbers significantly higher than those in randomised networks [1,3]. However, as already noted in [1], motifs that are functionally important but not statistically significant could exist and would be missed by this approach. Furthermore, statistically significant motifs are not necessarily functionally important.

The term *motif* is not precisely defined [1,2,3], but often refers to some functional properties of a (set of related) substructure(s) within a network or to their number of occurrences compared to randomised networks. To distinguish between substructures representing a motif and general substructures we use the term *pattern* for the latter. A pattern itself is defined as a single network. We allow the user to investigate all possible patterns in the network and do

C. Priami et al. (Eds.): Trans. on Comput. Syst. Biol. III, LNBI 3737, pp. 89–104, 2005.

not restrict our search to statistically significant substructures. This approach is similar to mining frequent patterns in networks as described in [5].

For the problem of pattern finding in collections of independent networks several algorithms have been presented [5,6,7]. However, there are only a few approaches to finding patterns in large connected networks [8,9]. This paper deals with two topics: 1) we introduce three different concepts for the determination of pattern frequency (the number of occurrences of a pattern in the target network) and 2) we present an algorithm to find frequent patterns of a given size under consideration of the different concepts for frequency counting. This algorithm is based on the approach presented in [8].

The remainder of the paper is structured as follows: in Sect. 2 we define the graph model on which we operate. Following this, in Sect. 3 we introduce three different concepts for the determination of pattern frequency. Section 4 presents the *flexible pattern finder* algorithm and Sect. 5 describes further enhancements of the algorithm. To demonstrate the utility of our method it is applied to typical real-world data from biology in Sect. 6. Furthermore, we investigate the execution time of a parallel version of the algorithm on a multiprocessor machine. Finally, we discuss our method and possible extensions in Sect. 7.

2   Definitions

We focus on directed graphs, however, the presented method works for both, directed and undirected graphs. A *directed graph* $G = (V, E)$ consists of a finite set V of vertices and a finite set $E \subseteq V \times V$ of edges. An edge $(u, v) \in E$ goes from vertex u, the source, to another vertex v, the target. The vertices u and v are said to be *incident* with the edge e and *adjacent* to each other. A subgraph of the graph $G = (V, E)$ is a graph $G_s = (V_s, E_s)$ where $V_s \subseteq V$ and $E_s \subseteq (V_s \times V_s) \cap E$. For a graph G and a subgraph G_s an edge $e = (u, v) \in E$ is called *incident* to G_s if exactly one of the vertices u, v is element of the set V_s. The *in-degree* of a vertex is defined as the number of edges coming into the vertex, the *out-degree* as the number of edges going out of it. The *degree* of a vertex is the number of all edges connected to it. A directed graph is called *connected* if it is possible to reach any vertex starting from any other vertex by traversing edges in some direction (i.e., for an edge (u, v) either the given direction from u to v or the direction from v to u). Two graphs $G_1 = (V_1, E_1)$ and $G_2 = (V_2, E_2)$ are *isomorphic* if there is a one-to-one correspondence between their vertices, and there is an edge directed from one vertex to another vertex of one graph if and only if there is an edge with the same direction between the corresponding vertices in the other graph.

Biological networks can be represented as graphs. For example, the proteins in a protein-protein-interaction network are modelled as vertices, the protein interactions are represented as edges. In a metabolic network the metabolites (compounds) and the reactions can be represented as vertices, and edges are binary relations connecting compounds of reactions with reaction vertices. Let G_T be a graph representing the biological network to be analysed (the *target*

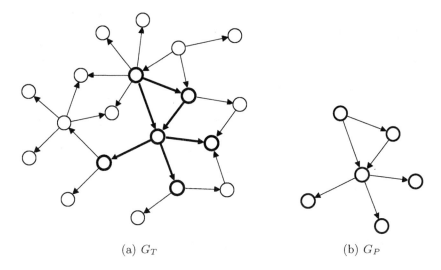

(a) G_T (b) G_P

Fig. 1. (a) A graph with a randomly selected subgraph (highlighted with bold lines). This subgraph is isomorphic to the graph G_P shown in (b). The highlighted subgraph in G_T is also a match of G_P in G_T.

network) and G_P be a connected graph representing the *pattern* of interest. A *match* G_M is a subgraph of G_T which is isomorphic to G_P (see Fig. 1). The *pattern size* is defined in this paper as the number of edges that the pattern comprises.

3   Pattern Frequency

3.1   Different Concepts for Determination of Pattern Frequency

The *frequency of a pattern* in a target graph is the maximum number of different matches of this pattern. There are different concepts for determining the frequency of a pattern depending on which elements of the graph can be shared by two matches. Tab. 1 lists the four possibilities, from which only three can be reasonably applied.

Frequency concept $\mathcal{F}_1$ counts every match of this pattern. This concept gives a complete overview of all possible occurrences of a pattern even if elements of the target graph have to be used several times. It does not exclude possible matches (as the other concepts usually do) and therefore shows the full 'potential' of the target graph with regard to the pattern.

Frequency concepts $\mathcal{F}_2$ and $\mathcal{F}_3$ restrict the reuse of graph elements shared by different matches of a pattern. If different matches share graph elements not allowed by the particular concept, not all matches can be counted for the frequency. In this case the maximum set of non-overlapping matches selected for frequency counting has to be calculated. This is known as the maximum

Table 1. Concepts for sharing of graph elements by the matches counted for the frequency of a pattern. Note that concept $\mathcal{F}^*$ is not applicable, since edges always connect vertices. In concept $\mathcal{F}_1$ where all elements can be shared, separate matches have to differ at least by one element. The concepts are applied to the graph and pattern in Fig. 2 and the results are shown at the right side of the table.

	Graph elements shared by different matches			Values for the example in Fig. 2	
Concept	Vertices	Edges	Frequency	Selected matches
$\mathcal{F}_1$	yes	yes	5	$\{M_1, M_2, M_3, M_4, M_5\}$
$\mathcal{F}_2$	yes	no	2	$\{M_1, M_4\}$ or $\{M_3, M_4\}$
$\mathcal{F}^*$	no	yes	–	–
$\mathcal{F}_3$	no	no	1	one of $\{M_1, M_2, M_3, M_4, M_5\}$

independent set problem. Concept $\mathcal{F}_2$ is a common concept [8,9] and only allows edge disjoint matches. Concept $\mathcal{F}_3$ is even more restrictive concerning sharing graph elements of the target graph for different matches. All matches have to be vertex and edge disjoint. One application is the study of processes in networks where each process (motif) consumes resources (vertices), which therefore can be used only once. The advantage of this concept is that the matches in the target graph can be seen as non-overlapping clusters. This clustering of the target graph allows specific analysis and navigation methods such as folding and unfolding of clusters and pattern preserving layout of the network.

The results of the application of the different concepts are illustrated by the example in Fig. 2. Applying concept $\mathcal{F}_1$, the frequency is five, for $\mathcal{F}_2$ the frequency is two counting the matches M_1 and M_4 or, alternatively M_3 and M_4. By applying concept $\mathcal{F}_3$ only one match out of the five can be selected.

3.2   Downward Closure Property of the Frequency

The downward closure property of the frequency of the patterns described in [8] allows the reduction of the search space for some frequency concepts and therefore a speed-up of the computation. The search algorithm traverses the patterns like a tree (see Sect. 4.2). The downward closure property ensures that the frequency of descending patterns (i.e., patterns on paths of the traversal tree) is monotonically decreasing with increasing size of the pattern. Based on this property, the search space of patterns can be reduced by pruning of infrequent patterns in the traversal tree. Intermediately discovered patterns which fall below a particular frequency threshold can be discarded together with all patterns descending from it, since no frequent patterns can be found in this branch of the traversal tree.

The downward closure property holds for frequency concepts $\mathcal{F}_2$ and $\mathcal{F}_3$ (see [8]), but not for $\mathcal{F}_1$. For $\mathcal{F}_1$ this is illustrated in Fig. 3 showing (a) a graph of size 9 and two patterns of (b) size 3 and (c) size 4. The pattern in (c) is a one-edge extension of the pattern in (b). The frequency of the pattern in (b) within the target graph in (a) is one for all three concepts. For the pattern in (c) the number of matches within the target graph in (a) is one for concept $\mathcal{F}_2$ and $\mathcal{F}_3$ and six for concept $\mathcal{F}_1$. In the latter case the number of matches increases for a pattern which is an extension of another pattern and hence is not downward closed.

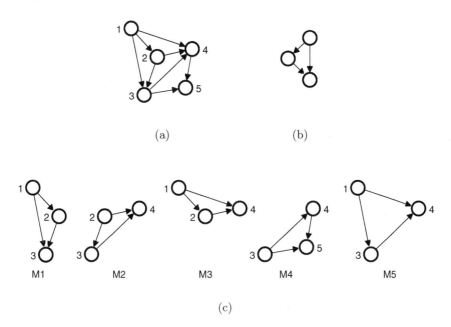

(a)   (b)

(c)

Fig. 2. An example graph (a), a pattern (b) and all different matches of the pattern (c, $M_1 - M_5$). The vertices of the graph and of the matches are numbered consecutively for identification purposes.

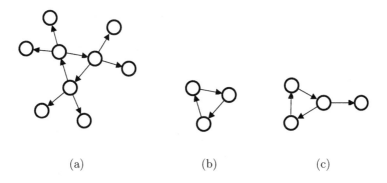

(a)   (b)   (c)

Fig. 3. A graph of size 9 (a) and two patterns of size 3 (b) and size 4 (c) illustrating that frequency concept $\mathcal{F}_1$ is not downward closed.

4   Flexible Pattern Finder Algorithm

4.1   Overview

In its basic form, the *flexible pattern finder* (FPF) algorithm searches for patterns of a given size (the target size) which occur with maximum frequency under a given frequency concept. The number of different patterns grows very quickly with increasing size of the patterns [10] (number of non-isomorphic graphs with n vertices: 1 ($n = 1$), 2 (2), 13 (3), 199 (4), 9364 (5), 1530843 (6)). Furthermore, there are up to $|E_t|^{|E_p|}$ ($|E_t|$ is the number of edges in the target graph and $|E_p|$ is the number of edges in the pattern) matches of pattern p in graph G_t for concept $\mathcal{F}_1$. Therefore, a systematic search of all patterns of a particular size can become very time consuming even for medium-size patterns. In order to avoid the generation of a high number of patterns FPF uses a method that builds a tree of only the patterns which are supported by the target graph and traverses this tree such that only promising branches are examined. To build this pattern tree, a particular pattern of size i is assigned to a parent pattern of size $i - 1$, the *generating parent*, from which it can be exclusively derived. Fig. 4 illustrates the concept.

For the frequency concepts $\mathcal{F}_2$ and $\mathcal{F}_3$ this pattern tree together with the downward closure property allows the pruning of the tree. As soon as the frequency of a pattern of intermediate size falls below the frequency of a pattern of target size discovered so far this branch of the tree can be discarded since the frequency of descending patterns is monotonically decreasing (see Sect. 3.2). If a (nearly) maximum frequent pattern of target size is discovered early in the

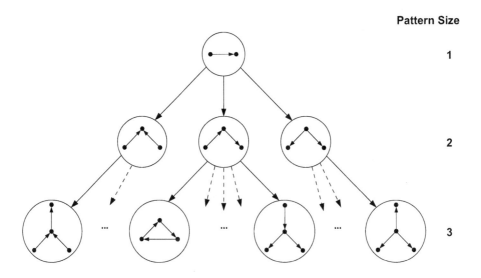

Pattern Size

Fig. 4. An example of a pattern tree. Note that there are more patterns of size 3 as indicated by dashed lines.

Algorithm 1: Flexible pattern finder

Data   : Graph $\mathcal{G}$, target pattern size t, frequency concept $\mathcal{F}$ (one of $\mathcal{F}_1$, $\mathcal{F}_2$, $\mathcal{F}_3$)
Result : Set $\mathcal{R}$ of patterns of size t with maximum frequency

$\mathcal{R} \longleftarrow \emptyset$; $f_{max} \longleftarrow 0$
/* $\mathcal{P}$: temporary data structure for patterns of intermediate size */
$\mathcal{P} \longleftarrow$ start pattern p_1 of size 1
$\mathcal{M}_{p_1} \longleftarrow$ all matches of p_1 in $\mathcal{G}$
while $\mathcal{P} \neq \emptyset$ **do**
    $\mathcal{P}_{max} \longleftarrow$ select all patterns from $\mathcal{P}$ with maximum size
    $p \longleftarrow$ select pattern with maximum frequency from $\mathcal{P}_{max}$
    $\mathcal{E} = \texttt{ExtensionLoop}(\mathcal{G}, p, \mathcal{M}_p)$
    foreach *pattern* $p \in \mathcal{E}$ **do**
        if $\mathcal{F} = \mathcal{F}_1$ **then**
            $f \longleftarrow size(\mathcal{M}_p)$
        else
            $f \longleftarrow \texttt{MaximumIndependentSet}\ (\mathcal{F}, \mathcal{M}_p)$ (see Sect. 4.5)
        end
        if $size(p) = t$ **then**
            if $f = f_{max}$ **then**
                $\mathcal{R} \longleftarrow \mathcal{R} \cup \{p\}$
            else
                if $f > f_{max}$ **then**
                    $\mathcal{R} \longleftarrow \{p\}$; $f_{max} \longleftarrow f$
                end
            end
        else
            if $\mathcal{F} = \mathcal{F}_1$ *or* $f \geq f_{max}$ **then**
                $\mathcal{P} \longleftarrow \mathcal{P} \cup \{p\}$
            end
        end
    end
end

---

search process, the frequency threshold for discarding intermediate size patterns becomes relatively high very early. Consequently, the number of patterns to be searched is reduced significantly. To find a frequent pattern early, the FPF algorithm selects the pattern with the highest frequency from the set of intermediate size patterns for extension.

In Sect. 4.2 the pattern extension and traversal of the flexible pattern finder algorithm are described, Sect. 4.3 to 4.5 contain details of the algorithm, and in Sect. 5 further enhancements are outlined.

4.2   Pattern Extension and Traversal

In general the patterns of intermediate size are extended until they reach the target size and then are logged as a result pattern or are discarded if they are

---

Algorithm 2: Extension Loop

    Data   : Graph $\mathcal{G}$, pattern p of size i, set of all matches $\mathcal{M}_p$ of pattern p
    Result : Set of extension patterns $\mathcal{E}$ of size $i + 1$ together with all matches for
             each valid extension pattern

    $\mathcal{E} \longleftarrow \emptyset$
    foreach *match* $m \in \mathcal{M}_p$ **do**
        foreach *incident edge*[1] e *of* m
        do
            $m' \longleftarrow m \cup e$
            $p' \longleftarrow$ corresponding pattern of m'
            if p *is generating parent of* p' **then**
               $\mathcal{E} \longleftarrow \mathcal{E} \cup \{p'\}$; $\mathcal{M}_{p'} \longleftarrow \mathcal{M}_{p'} \cup \{m'\}$
            end
        end
    end

---

infrequent. The search process terminates if no patterns of intermediate size are left for extension.

The search starts with the smallest pattern p of size 1 which consists of one edge and two vertices. All matches $\mathcal{M}_{p_1}$ of this pattern are generated using every edge once. Matches are stored as sets of edges. On the basis of the matches $m \in \mathcal{M}_p$ of a pattern p of size i all one-edge extension patterns p' of size $i+1$ are created, restricted to patterns which have p as generating parent. This is done in the following way: for every match m all incident edges within the graph are identified and are used successively to create a new match m' of size $i + 1$. The resulting extension pattern p' of m' is discarded if p is not the generating parent of p', see Sect. 4.3. If it is a valid extension, it is checked whether this pattern was discovered before using a canonical labelling for isomorphism testing, and it is recorded as an extension pattern if not already present. The matches $\mathcal{M}_{p'}$ for all newly created extension patterns are identified during the process of pattern generation. These newly created extension matches are recorded and are assigned to the matches of the corresponding extension pattern. The complete algorithm is shown in Alg. 1 and Alg. 2.

4.3   Generating Parent

A pattern of size i can be derived from up to i different patterns of size $i - 1$ by adding one edge. In order to avoid the redundant generation of a pattern only one defined pattern of size $i - 1$ is allowed to generate this pattern, the *generating parent*. It is defined on the basis of the canonical ordering (see Sect. 4.4) of the adjacency matrix of the graph. The last edge defined by the top-to-bottom and left-to-right ordering of the adjacency matrix which does not disconnect the remaining edges of the pattern is removed. The remaining size $i - 1$ pattern is the generating parent of the given size i pattern.

---

[1] See definition of edge between graph and subgraph.

4.4   Canonical Label

The canonical label is a string representing a unique identifier of a graph. This label is based on the adjacency matrix representation of the graph. The adjacency matrix is ordered in a defined way, so that the label is invariant to the initial ordering of the matrix. By comparing the canonical labels graphs can be checked for isomorphism. The principle of the algorithm for the generation of a canonical label is described in [11]. Some modifications are made with respect to different properties of the input data.

4.5   Maximum Independent Set

Whereas the frequency of a pattern in concept $\mathcal{F}_1$ is given by all matches of this pattern in the target graph, for the concepts $\mathcal{F}_2$ and $\mathcal{F}_3$ the pattern frequency is calculated after the determination of all matches by computing the maximum independent set on the basis of their overlap graph. The overlap graph is constructed by inserting a vertex for each match of the pattern. An edge is inserted between a pair of vertices if their corresponding pattern matches overlap, i.e., they share graph elements which the particular concept does not allow for different matches counted for the frequency. The maximum independent set S of a graph $G = (V, E)$ is defined as the largest subset of vertices of V such that no pair of vertices of S defines an edge of E. Since the problem of finding the maximum independent set is $\mathcal{NP}$-complete [12], a heuristic can be used to approximate the exact result for performance reasons. However, it should be noted that by using a heuristic the frequencies computed by the algorithm are only lower bounds, furthermore the algorithm may not find all patterns with highest frequency.

5   Enhancements of the Search Process

5.1   Sets of Frequent Patterns

The presented algorithm searches for patterns of a given size which occur with maximum frequency under a given frequency concept. Usually not only these patterns are of interest, but all patterns with a frequency above a given threshold or a set of patterns with highest frequency where the size of the set is restricted. In both cases pruning of the search tree is possible for frequency concepts $\mathcal{F}_2$ and $\mathcal{F}_3$.

In the first case a given frequency threshold f_t is used as a global variable for discarding infrequent patterns during the whole search process. In Alg. 1 f_t is assigned to f_{max} during the initialisation of the algorithm. Furthermore, f_{max} is not changed if the frequency of the current pattern is higher than f_{max}. If the global threshold is set to 1 all patterns supported by the target graph are searched.

The second case is searching for at least n patterns with highest frequency. Here the frequency of the maximum frequent pattern computed so far is not used as the frequency threshold f_{max} for discarding infrequent patterns, rather f_{max} is given by the current minimum of the frequencies of the n most frequent patterns. Note that this extension may compute more than n highly frequent patterns.

5.2   Parallel Processing

As each pattern has one generating parent and the branches of the pattern tree can be traversed independently of each other the algorithm can be processed in parallel on multiprocessor machines to speed-up the overall search process. For pruning of infrequent patterns the current frequency threshold f_{max} has to be communicated between the parallel tasks. In cases when no pruning of the search tree is possible, either because frequency concept $\mathcal{F}_1$ is applied or a global threshold for the minimum frequency exists which does not change during the search process, even this communication is obsolete. The extensions to run the algorithm on parallel computers are not shown in Alg. 1 and Alg. 2, however, in Sect. 6.3 the speed-up of the computation using the parallel implementation is analysed on a multiprocessor machine using typical real-world data.

5.3   Pattern Size

The size of patterns is given by the number of edges. This is motivated by the way the FPF algorithm extends patterns. However, it is common to define the size of a pattern by its number of vertices [1,2,3]. To extend the flexible pattern finder algorithm to search for patterns where the pattern size is defined as the number of vertices, two changes to the algorithm are necessary: the computation of the set $\mathcal{P}$ of patterns to be extended and a second method for pattern extension. Usually, when a pattern has reached the target size (the number of edges in the original algorithm) this pattern will not be further extended and will not be included into $\mathcal{P}$. Now patterns have to be extended if the size of the new pattern (that is now the number of vertices) does not increase, which happens when a new edge between two already existing vertices is included into the pattern. If $size(p) = t$ in Alg. 1 patterns which are recorded as result patterns $\mathcal{R}$ are also added to $\mathcal{P}$. The second method for pattern extension changes Alg. 2. For patterns of target size only edges between two already existing vertices are selected for extensions.

6   Results

6.1   Application to Transcriptional Regulatory Network

As one application of our algorithm we took a data set of interactions of transcription factors in *Saccharomyces cerevisiae*. It was published in [13] and is available under http://jura.wi.mit.edu/young_public/regulatory_network/. It comprises 106 transcriptional regulators with 108 interactions, where an inter-

Table 2. Patterns with highest frequency for the different concepts for frequency counting, in (a) for concept $\mathcal{F}_1$, in (b) for $\mathcal{F}_2$ and in (c) for $\mathcal{F}_3$. Since in (c) 30 patterns occur with the maximum frequency of five for concept $\mathcal{F}_3$, the three patterns which have maximum respective minimum frequency for the other two concepts were chosen from these 30 patterns. In all cases the frequencies of the patterns for the other two concepts are also presented. The patterns are shown in Fig. 5.

<table>
<tr><td colspan="4" align="center">(a)</td><td colspan="4" align="center">(b)</td><td colspan="4" align="center">(c)</td></tr>
<tr><td>Pattern</td><td colspan="3">Frequency</td><td>Pattern</td><td colspan="3">Frequency</td><td>Pattern</td><td colspan="3">Frequency</td></tr>
<tr><td></td><td>$\mathcal{F}_1$</td><td>$\mathcal{F}_2$</td><td>$\mathcal{F}_3$</td><td></td><td>$\mathcal{F}_2$</td><td>$\mathcal{F}_1$</td><td>$\mathcal{F}_3$</td><td></td><td>$\mathcal{F}_3$</td><td>$\mathcal{F}_1$</td><td>$\mathcal{F}_2$</td></tr>
<tr><td>P_1</td><td>3175</td><td>6</td><td>3</td><td>P_4</td><td>12</td><td>1627</td><td>5</td><td>P_2</td><td>5</td><td>2785</td><td>10</td></tr>
<tr><td>P_2</td><td>2785</td><td>10</td><td>5</td><td>P_5</td><td>11</td><td>1194</td><td>5</td><td>P_4</td><td>5</td><td>1627</td><td>12</td></tr>
<tr><td>P_3</td><td>2544</td><td>9</td><td>4</td><td>P_6</td><td>11</td><td>996</td><td>5</td><td>P_7</td><td>5</td><td>303</td><td>5</td></tr>
</table>

action stands for the binding of one factor to the promoter of the gene of a regulated factor and is therefore directed.

Searching for all patterns with 6 edges resulted in 1811 non-isomorphic patterns covering a broad range of frequency. Table 2 lists the patterns with highest frequency for the different concepts of frequency counting. Figure 8 shows a part of the network with the matches of pattern P_2 (see Tab. 2) under consideration of frequency concept $\mathcal{F}_3$.

6.2   Application to Metabolite Network

For the second application of the FPF algorithm and for the analysis of the parallel execution (Sect. 6.3) we took metabolite networks published in [14]. The data were derived from the *KEGG LIGAND Database* and were revised by the authors, which also added reversibility information to the reactions. The network represents the connectivity of metabolites established by reactions, where the metabolites represent the vertices and edges between vertices represent the conversion of one metabolite into the other performed in a reaction. If the reaction is irreversible the edge is directed from the substrate of the reaction to the product. In case of a reversible reaction, an additional edge in the reverse direction is present.

We analysed the human (*Homo sapiens*) metabolite network which comprises 1054 vertices and 1717 edges. A total of 3958 non-isomorphic patterns with 5 vertices were found. Table 3 lists the patterns with highest frequency for the different concepts of frequency counting. Comparing the structure of the patterns with highest frequency (see Fig. 6) shows that it is somewhat uniform for each concept but different between $\mathcal{F}_1$ to $\mathcal{F}_2$ and $\mathcal{F}_3$. For $\mathcal{F}_1$ all three patterns have four vertices only connected to the central vertex. For $\mathcal{F}_2$ and $\mathcal{F}_3$, which partially share the same patterns, all patterns are linear.

6.3   Parallel Execution of the FPF Algorithm

We implemented the described enhancements (see Sect. 5) of the FPF algorithm and did some experiments to analyse the speed-up on a parallel computer. The

Table 3. The three patterns with highest frequency for each frequency concept of the human metabolite network in Sect. 6.2. The patterns are shown in Fig. 6. Note that in this example the size of a pattern is given by the number of vertices.

(a)					(b)					(c)			
Pattern	Frequency				Pattern	Frequency				Pattern	Frequency		
	$\mathcal{F}_1$	$\mathcal{F}_2$	$\mathcal{F}_3$			$\mathcal{F}_2$	$\mathcal{F}_1$	$\mathcal{F}_3$			$\mathcal{F}_3$	$\mathcal{F}_1$	$\mathcal{F}_2$
P_a	30931	140	58		P_d	291	15586	93		P_g	93	16730	269
P_b	24967	117	52		P_e	283	14661	84		P_d	93	15586	291
P_c	24249	104	50		P_f	280	15349	92		P_f	92	15349	280

algorithm is implemented in Java. The system used for the analysis was a Sun Fire V880, equipped with eight 1.2 GHz UltraSPARC III processors each having 2 GB RAM. The operating system was SunOS Release 5.9 and for Java the Java HotSpot(TM) 64-Bit Server Virtual Machine (build 1.4.2_06-b03) was employed. The Sun Fire V880 supports symmetric multiprocessing. The execution time to search for all patterns with 5 and 6 vertices for different numbers of processors was measured for 10 different metabolite networks.

Generally, the execution time decreases for a particular network with increasing number of available processors, see Fig. 7. However the speed-up is somewhat less than the number of used processors. This is caused by the overhead of working with different threads for parallel execution and, with a greater effect, by the traversal of the pattern tree. It cannot be guaranteed that enough branches for independent pattern traversal are present at any time to employ all available processors, because the pattern tree depends on the supported patterns of the

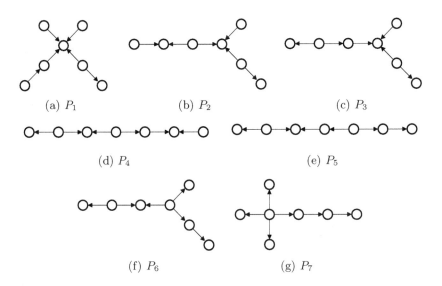

(a) P_1 (b) P_2 (c) P_3

(d) P_4 (e) P_5

(f) P_6 (g) P_7

Fig. 5. The frequent patterns in Tab. 2

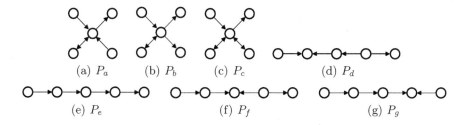

(a) P_a (b) P_b (c) P_c (d) P_d

(e) P_e (f) P_f (g) P_g

Fig. 6. The frequent patterns in Tab. 3

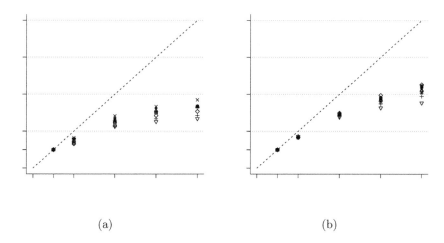

(a)   (b)

Fig. 7. Speed-up of the pattern search algorithm on a multiprocessor machine for different numbers of processors. Ten different metabolite networks (see Sect. 6.2) with size greater than 590 vertices were searched for all patterns with (a) 5 and (b) 6 vertices. The execution time was determined for each network and each number of processors based on the mean value of three passes.

particular network. Comparing the execution times for the search for patterns with 5 and 6 vertices shows a lesser variance for the execution times for the different networks in case of greater patterns. Furthermore, the speed-up in case of larger patterns seems to be higher than the speed-up for smaller patterns. Larger pattern size leads to a higher number of different patterns. The higher number of patterns results in a better working load of the available processors due to a bigger traversal tree, as mentioned before.

7   Discussion

This paper discussed two topics related to network-motif detection and frequent pattern mining: 1) three different concepts for the determination of pattern fre-

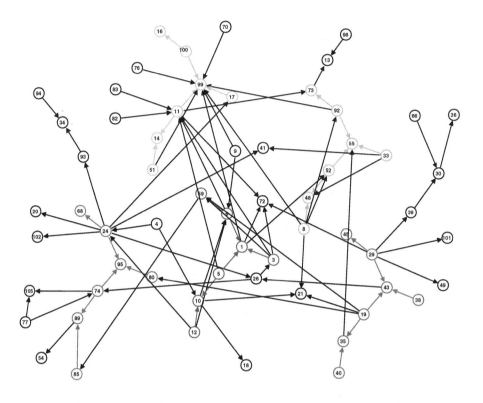

Fig. 8. A part of the network described in Sect. 6.1 with the matches of pattern P_2 (see Tab. 2) under consideration of frequency concept $\mathcal{F}_3$. The matches are highlighted with different colours and are similarly layouted.

quency and 2) a flexible algorithm to compute these frequencies. Analysing networks and finding interesting patterns is not only important in biology as shown, but has applications in other fields of science such as the toxicology or carcinogenicity of chemical substances [15] and the classification of sequential logic circuits [1].

Different frequency concepts may be used to focus on particular questions. To reveal all possible occurrences of a specific pattern, frequency concept $\mathcal{F}_1$ provides the desired information. Frequency concepts $\mathcal{F}_2$ and $\mathcal{F}_3$ are appropriate for example if the maximum number of instances of a particular pattern which can be "active" at the same time is demanded or if non-overlapping matches for visualisation purposes are of interest. The flexible pattern finder algorithm is an extension of [8] to find frequent patterns of a given size under consideration of the different frequency concepts. Several enhancements have been presented and a parallel implementation of the algorithm has been studied.

The algorithm has been applied to two different biological networks, a transcriptional regulatory network and a metabolite network. Remarkably, the values for frequency concepts $\mathcal{F}_1$ in comparison to $\mathcal{F}_2$ and $\mathcal{F}_3$ are up to two orders of magnitude higher for frequent patterns. Comparing the values for the different concepts shows that a high frequency of a specific pattern for one concept does not necessarily imply a high frequency for another concept. The frequency of a pattern alone is not sufficient for identifying functional motifs. Whether a frequent pattern has a functional role in the network has to be analysed further. Therefore this step is only a starting point which generates interesting hypotheses for further studies.

The algorithm considers only the topology of the network, but does not take additional information about the involved elements and relations into account. This is an abstraction of the functional aspects of the biological network, however it keeps the major relationships of the elements forming the network. The incorporation of more detailed information such as edge weights or vertex labels into the search is planned for further development. As the parallel version of the algorithm has just been implemented, the further analysis of its scalability and improvements are the next step. Finally, we plan to extend the visual analysis of the patterns in the target network by developing specific graph layout methods for this task.

Acknowledgements

This work was supported by the German Ministry of Education and Research (BMBF) under grant 0312706A. We would like to thank Christian Klukas for providing an implementation of the pattern-preserving force-directed layout algorithm, the group of Franz J. Brandenburg (University of Passau, Germany) for the excellent cooperation and for granting usage of Gravisto which was used for implementing the algorithm and the reviewers for their helpful comments.

References

1. Milo, R., Shen-Orr, S., Itzkovitz, S., Kashtan, N., Chklovskii, D., Alon, U.: Network motifs: Simple building blocks of complex networks. Science **298** (2002) 824–827
2. Shen-Orr, S., Milo, R., Mangan, S., Alon, U.: Network motifs in the transcriptional regulation network of Escherichia coli. Nature Genetices **31** (2002) 64–68
3. Wuchty, S., Oltvai, Z.N., Barabási, A.L.: Evolutionary conservation of motif constituents in the yeast protein interaction network. Nature Genetics **35** (2003) 176–179
4. Mangan, S., Alon, U.: Structure and function of the feed-forward loop network motif. Proceedings of the National Academy of Sciences **100** (2003) 11980–11985
5. Inokuchi, A., Washio, T., Motoda, H.: Complete mining of frequent patterns from graphs: Mining graph data. Machine Learning **50** (2003) 321–354
6. Kuramochi, M., Karypis, G.: Frequent subgraph discovery. In: IEEE International Conference on Data Mining (ICDM). (2001) 313–320
7. Yan, X., Han, J.: gspan: Graph-based substructure pattern mining. In: IEEE International Conference on Data Mining (ICDM). (2002) 721–724

8. Kuramochi, M., Karypis, G.: Finding frequent patterns in a large sparse graph. In: SIAM International Conference on Data Mining (SDM-04). (2004)

9. Vanetik, N., Gudes, E., Shimony, S.E.: Computing frequent graph patterns from semistructured data. In: IEEE International Conference on Data Mining (ICDM). (2002) 458–465

10. Harary, F., Palmer, E.M.: Graphical Enumeration. Academic Press, New York (1973)

11. Kuramochi, M., Karypis, G.: An efficient algorithm for discovering frequent subgraphs. Technical Report 02-026, Department of Computer Science, University of Minnesota (2002)

12. Garey, M., Johnson, D.: Computers and Intractability: A Guide to the Theory of NP-Completeness. W.H. Freeman and Company, New York (1979)

13. Lee, T.I., Rinaldi, N.J., Robert, F., Odom, D.T., Bar-Joseph, Z., Gerber, G.K., Hannett, N.M., Harbison, C.T., Thompson, C.M., Simon, I., Zeitlinger, J., Jennings, E.G., Murray, H.L., Gordon, D.B., Ren, B., Wyrick, J.J., Tagne, J.B., Volkert, T.L., Fraenkel, E., Gifford, D.K., Young, R.A.: Transcriptional regulatory networks in Saccharomyces cerevisiae. Science **298** (2002) 799–804

14. Ma, H., Zeng, A.P.: Reconstruction of metabolic networks from genome data and analysis of their global structure for various organisms. Bioinformatics **19** (2003) 270–277

15. Srinivasan, A., King, R.D., Muggleton, S.H., Sternberg, M.J.E.: The predictive toxicology evaluation challenge. In: 15th International Joint Conference on Artificial Intelligence (IJCAI). (1997) 1–6

An Agent-Oriented Conceptual Framework
for Systems Biology

Nicola Cannata[1,*], Flavio Corradini[2], Emanuela Merelli[2,**],
Andrea Omicini[3], and Alessandro Ricci[3]

[1] CRIBI Biotechnology Centre, Università di Padova,
via Ugo Bassi 58/B, 35131 Padova, Italy
nicola@cribi.unipd.it
[2] Dipartimento di Matematica e Informatica, Università di Camerino,
Via Madonna delle Carceri 62032 Camerino, Italy
{flavio.corradini, emanuela.merelli}@unicam.it
[3] DEIS, Alma Mater Studiorum, Università di Bologna,
Via Venezia 52, 47023 Cesena, Italy
{andrea.omicini, a.ricci}@unibo.it

Abstract. Recently, a collective effort from multiple research areas has
been made to understand biological systems at the system level. On the
one hand, for instance, researchers working on *systems biology* aim at understanding how living systems routinely perform complex tasks. On the
other hand, bioscientists involved in pharmacogenomics strive to study
how an individual's genetic inheritance affects the body's response to
drugs. Among the many things, research in the above disciplines requires
the ability to simulate particular biological systems as cells, organs, organisms and communities. When observed according to the perspective
of system simulation, biological systems are complex ones, and consist
of a set of components interacting with each other and with an external
(dynamic) environment.

In this work, we propose an alternative way to specify and model
complex systems based on behavioral modelling. We consider a biological system as a set of active computational components interacting in a
dynamic and often unpredictable environment. Then, we propose a *conceptual framework* for engineering computational systems simulating the
behaviour of biological systems, and modelling them in terms of *agents*
and *agent societies*.

1 Introduction

In the last years, the mutual exchange of ideas between researchers from biology and computer science has resulted in the emergence of new research fields
such as bioinformatics, computational biology and, more recently, systems biology. While the terms 'bioinformatics' and 'computational biology' have quite a
blurred acceptation, and are often used in an interchangeable way, systems biology aims at system-level understanding of biological systems [1]. So, the term

---

* Supported by grant GSP04289 from the Italian Telethon Foundation.
** Work completed while the author was on Fulbright leave at the University of Oregon.

C. Priami et al. (Eds.): Trans. on Comput. Syst. Biol. III, LNBI 3737, pp. 105–122, 2005.
© Springer-Verlag Berlin Heidelberg 2005

'systems biology' is typically used to indicate the attempt to integrate the huge and multiform amount of biological data in order to understand the behaviour of biological systems, and to study the relationships and interactions between the various parts of a biological system, such as organelles, cells, physiological systems, organisms etc.

The rapid growing of knowledge at the molecular level (i.e. genomes, transcriptomes, proteomes, metabolomes) is giving, for the first time, the opportunity to constitute the solid ground upon which to create an understanding of living organisms at the system level. Such an effort is meant not only at describing in detail the system structure and behaviour, but also at understanding its reaction in response to external stimuli or disruptions, as in the Systeome project [1]. Examples of well-studied and understood molecular networks include gene regulation networks (how genes and their products, proteins and RNAs, are regulating gene expression), metabolic pathways (the network of reactions involving metabolites) and signal transduction cascades (the molecular interactions activating a genetic answer to a signal received from the cell).

Because of the scale, nature and structure of the data, this new scientific challenge is much more demanding than ever, and is going to involve computer scientists and engineers, mathematicians, physicists, biochemists, and automatic control systems experts, working in close partnership with life scientists. Along this line, fundamental issues are the information management framework, as well as the model construction, analysis and validation phases [2]. Efforts are ongoing, meant to provide for a common and versatile software platform for systems biology research. For instance, in the Systems Biology Workbench project [3], some of the critical issues regard the exchange of data and the interface between software modules: SBML [4, 5] and CELLML [6] are modelling languages in systems biology that are aimed at facing such issues.

Although systems biology has no clear end-point, the prize to be attained is immense. From in-silico drug design and testing, to individualised medicine, which will take into account physiological and genetic profiles, there is a huge potential to profoundly affect health care, and medical science in general [2].

Therefore, in this article we propose an agent-oriented conceptual framework to support the modelling and analysis of biological systems. The framework is intended to support life scientists in building models and verifying experimental hypotheses by simulation. We believe that the use of an agent-based computational platform [7, 8] and agent coordination infrastructures [9], along with the adoption of formal methods [10] will make it possible to harness the complexity of the biological domain by delegating software agents to simulate bio-entities [11, 12, 13]. Also, by supporting the integration of different models and tools [14], an agent-oriented framework could allow life scientists to analyse and validate system behaviours [15], to verify system's properties [16], and to design, in a compositional way, new models from known ones. The conceptual framework we propose takes into account the four steps suggested by Kitano in [1]: (i) system structure identification, (ii) system behaviour analysis, (iii) system control, (iv) system design. For each step, our framework exploits agent-oriented metaphors, models and infrastructures to provide systems biologists with the suitable methodologies and technologies.

Accordingly, in the following Section 2 describes the steps required in order to understand biological systems. Then, Section 3 introduces the main technologies used in the systems-biology field in relation to the conceptual framework cited above. Afterwards, Section 4 motivates the need for an agent-oriented framework in systems biology, and Section 5 points out the role of agents and define the conceptual framework, which is exemplified in Section 6. Finally, Section 7 concludes the paper and discusses the possible future directions of our work.

2   Understanding Biological Systems

A biological system is an assembly of inter-related biological components. Besides describing its components in details, in order to understand a biological system it is also necessary to describe the behaviour of any component at the molecular level, and to comprehend what happens when certain stimuli or malfunction occurs. To promote system-level understanding of biological systems, Kitano suggests to investigate along the following lines [1].

- First of all, we need to identify the *structure* of the biological system, which leads to the specification of the topological relationship of the network of components, as well as the parameters for each relation. In this stage, all the possible experimental data should be used to fulfil this purpose. For example, to specify a metabolic pathway, one should first identify all components of the pathway, the function of each component, the interactions among them and all the associated parameters; then characterise any system component with the biological knowledge provided by the research in molecular biology.
- Then, we need to understand the system *behaviour*, that is, the mechanisms that are behind the robustness and stability of the system, and the functionalities of the interactions among components. *Simulation* is an essential tool to understand the behaviour of biological systems. For example, to simulate a metabolic pathway one should take raw biological data, then choose a formal notation and model, like direct graphs, Bayesian networks, Boolean networks, ordinary and partial equations, qualitative differential equations, stochastic equations or rule-based formalism [17].
- By the system *control* mechanisms, we can then increase the stability of the system during the simulation of a complex system. In particular, in biological systems the two most extensively-used control schemas are the feedforward and the feedback control. The feedforward control represents an open-loop control where a set of predefined reactions sequences is triggered when a certain stimulus is present. The feedback control is a close-loop control which allows the signal in output to be processed as one of the input of the system, and therefore behaviour of the system to be controlled as desired. Many examples could be found which demonstrate the usefulness of feedback control—among which the control of the growing process of human cells against the cancer.

 – Finally, we must be able to *design* biological systems for providing cure and
   supporting pharmacogenomics, i.e., the study of how an individual's genetic
   inheritance affects the body's response to drugs.

System's modelling, simulation and analysis are activities that typically imply the
choice and the integration of more than one model and more than one computational tool. For instance, a bioscientist may want to retrieve raw data coming from
public databases, mix and match several models (qualitative and quantitative) on
that data, then simulate and analyse the behaviour of the resulting system—and
in some cases also produce large datasets to be afterwards analysed [18].

   Thus, a framework that would integrate, in a consistent way, a number of different computational components could be fundamental in assisting researchers
in systems biology. Languages and compilers to model and store biological systems, wrappers to access and collect experimental data from different databases,
biological systems simulators and analysis tools to support the behavioural
analysis—these could be some of the components characterising the proposed
conceptual framework. So, in general, we aim at defining a conceptual framework which allows biologists, given the structure and behaviour of a biological
system as input, to create the corresponding model and to simulate the system
by studying its behaviour and by verifying the properties of each component. Of
course, a more sophisticated future scenario is that in which a biologist does not
know the system structure and behaviour (i.e., he/she only can access a huge
amount of data), but anyway aims at inferring a new model by mining biological
data and knowledge dispersed all over the world.

3   Related Works

In the literature, a number of different approaches to biological systems simulation can be found, along with some, more or less limited, successful modelling
examples.

Ordinary Differential Equations. This is the classical approach arising from
the biochemical point of view. A network of interactions (chemical reactions)
between molecules (metabolites, proteins, genes) is established and Ordinary
Differential Equations (ODE) are used to numerically describe the continuous
variations in the concentration of substances. The GEPASI program [19], even
with a limited number of metabolites and reactions is for sure a milestone in
this area. The group of Mendes at the Virginia Bioinformatics Institute, in collaboration with the Kummer group at EML is currently accomplishing the development of COPASI, the successor of GEPASI, capable of carrying out more
sophisticated analysis (stochastic integration, non-linear dynamic analysis such
as bifurcation). Another ODE-based software environment is E-CELL [20], in
which the user can define protein functions, protein-protein and protein-DNA
interactions and regulation of gene expression, then observe the dynamics of the
changes in the concentrations of proteins, protein complexes and other chemical
compound in the cell. The authors simulated also a hypothetical cell with a 127
genes genome sufficient for the processes of transcription, translation, energy

production and phospholipid synthesis. Last but not least, we mention Simpathica and the related XS-system [21] which allow the user to describe and interact with biochemical reactions. Using a simple graphical or textual representation like SBML (Systems Biology Markup Language) or MDL (Model Description Language), a mathematical description in terms of differential algebraic description, ordinary, probabilistic, or stochastic differential equations is implemented. The system supports a wide range of analysis tools: model checking with a propositional branching-time temporal logic, time-frequency analysis tools with wavelets, linear discriminant bases, Karhunen-Loeve analysis, clustering using wavelet-bases, information-theoretic approaches to measure the significance of the system components, and approaches based on statistical learning theory.

Petri Nets. Many of the existing approaches for the modelling and simulation of biological systems lack effective and intuitive GUI interfaces. The use of an architecture based on Petri Nets addresses this problem, because of their intuitive graphical representation and their capabilities for mathematical analysis. Several enhanced Petri Nets (for example coloured Petri nets, stochastic and hybrid Petri Nets) have been used to model biological phenomena [22, 23, 24] but a more suitable approach is constituted by hybrid Petri nets that take in account both the discrete and continuous dynamics aspects [25].

Process Calculi. This approach provides textual languages close to convectional expressions used in biology to describe system structure. These languages are concurrent and compositional, and make it possible to concentrate on specific aspects of the systems biology. They are suitable to harness the complexity of the systems themselves. In Bio-calculus [26] a bio-syntax and multi-level semantics of systems are provided to describe and simulate some molecular interactions. Regev et al. use π-calculus to describe biomolecular processes [27, 28]. This process algebra has been originally introduced to describe computer processes and a simulation system (BioSPI) for the execution and analysis of a π-calculus program developed and experimented with a model for the RTK-MAPK signal transduction pathway. Another example of π-calculus application is the VICE virtual cell [29]. The authors observe that the cell mechanisms and global computing applications are closely related, and that biological components can be thought as processes while organisms as networks. The interactions between biological components are then represented by the communications between processes. They also proposed and obtained a very basic prokaryote-like genome which contains 180 genes: for this hypothetic organism, they experimented the essentials metabolic pathways. In [30, 10] a promising notational framework for unifying different aspects of biological representation is presented. The proposed process calculi framework is suitable for relating different levels of abstractions, which is going to be essential for feasibly representing biological systems of high architectural complexity.

Both Petri Nets and process algebra allow biological systems to be modelled by a formal specification, which is suitable to system property verification and behaviour analysis [31]. The need for a formalism to naturally describe the static-structure semantics of the biological system is also emphasised by Peleg et al. in [22] with respect to Petri Nets.

Multi-Agent Systems. Rather than an exclusive alternative to process algebra or Petri Nets, we consider the agent-based approach as a suitable approach to the engineering of simulation systems by exploiting also the semantics description of the biological system. As an example, the Cellulat model is the combination of the behavioural-based paradigm and the blackboard architecture for intracellular signalling modelling [32]. Also d'Inverno and Saunders [12] adopt a multiagent system to model and simulate the behaviour of stem cells. Both papers show specific and concrete examples of what could be done with an agent-oriented conceptual approach to cell and molecular biology.

Thus, beside the potential benefits of using process algebras to understand complex systems by decomposing them into simpler subsystems, analyse properties of subsystems using established process calculi theory, predict behaviour of subsystems by running stochastic simulations and predict properties and behaviour of composed systems, the agent-oriented approach can also support the modelling of static-structural system properties.

4 Motivation

From the previous section, the need for formal frameworks for modelling, simulating and analysing biological systems clearly emerges. Universally accepted and computer-understandable formal languages are needed, designed to describe complex regulation systems of biochemical processes in living cells. It has been observed [33] that the language used by biologists for verbal communications is vague and avoid clear prediction. Also, the use of colourful diagrams, so much beloved from biologists, is often usually useless for computer reasoning and quantitative analysis. Formal techniques are instead likely to give the chance to inherit the wide range of existing methods and tools provided by computer science (such as property design and verification, and automated reasoning).

Nevertheless, an important issue concerns the intuitive graphical representation of systems, along with the network of all the interactions among the entities, and the possibility to define it in an easy way through an user friendly GUI. Therefore, we should focus on a framework that allows for an easy "translation" from a graphical description of a system into a formal notation describing it. The choice of the computational paradigm also affects the possibility to perform more complicated tasks like stochastic integration or non-linear dynamic analysis, and to describe discrete and continuous hybrid systems.

What we found illuminating was the "molecule-as-computation" abstraction presented in [34], in which a system of interacting molecular entities is described and modelled by a system of interacting computational entities. Abstract computer languages, originally developed for the specification and study of systems of interacting computations are now actively used to represent biomolecular systems, including regulatory, metabolic and signalling pathways as well as multi-cellular processes. *Processes*, the basic interacting computational entities of these languages have an internal state and interaction capabilities. Process *behaviour* is governed by reaction rules specifying the response to an input message based on its content and on the state of the process. The response can include state change, a change in interaction capabilities, and/or the sending of messages.

Complex entities are described hierarchically. So, using the process abstraction opens up new possibilities for understanding molecular systems. Computers and biomolecular systems both start from a small set of elementary components from which, layer by layer, more complex entities are constructed with evermore sophisticated functions. While computers are networked to perform larger and larger computations, cells form multicellular organs and organisms, and organisms build societies.

The framework should also allow the system to be zoomed in and out at different levels of abstraction. As we cannot, and need not, to recreate the world as an isomorphic in-silico image of itself [2], it makes no sense to start the modelling from the atomic level. At the molecular level we can consider the cell chemical components: water, inorganic ions, sugars, aminoacids, nucleotides, fatty acid and other small molecules. These components can interact with each others and can be used to build up more complex macromolecules—like polysaccharides, composed of sugars, nucleic acids (DNA and RNA), composed of nucleotides, and proteins, composed of aminoacids. Macromolecules can then generate macromolecular aggregates—for example, the ribosome is made out of proteins and RNAs. Molecules can have an internal state—for instance, proteins can have different conformational states, they can be phosphorilated, the DNA can be methylated and the RNA can form secondary structure. Another level of modularity can be found in proteins domains (autonomous functional subunit of the proteins) and in nucleic acids "signals", for example transcription factor binding sites in the DNA, protein binding sites in the RNA, which usually are involved in the molecular interactions. The interactions at the various levels are often modelled by chemical bonds, either very strong and stable (for example the peptidic bond between aminoacids) or weak and temporary (for example the binding of proteins to DNA or between proteins). The composition of large number of simpler interactions makes up the cellular processes that are fundamental for the physiology of the cell—for example, the DNA replication (to propagate the genetic inheritance to the following generations) and the gene expression, which is composed of more phases, the most important being the transcription of DNA into RNA and the the translation of mRNA into protein (to synthesize the molecular machinery necessary for the normal and for the perturbed life of the cell). With the fast growth of biomolecular knowledge consequent to the -omics revolutions, for each of these processes it would soon be possible to individuate which are the actors involved, which is their role, which are the interactions between them, which is the result of the interactions, and which new entities are produced by the composition of simpler entities.

5   An Agent-oriented Framework for Systems Biology

Generally speaking, multiagent systems (MASs) are considered a suitable framework for modelling and engineering complex systems, characterised by organisation structures and coordination processes that are more and more articulated and dynamic [35, 36]. Also, MASs are considered as a promising approach for engineering simulations of complex systems, as one can see from the series of the

Multi-Agent Based Simulation (MABS) workshops—held since 1998—or from the Journal of Artificial Societies and Social Simulation (JASSS).

In particular, MAS-based models are often used for the simulation of systemic and social phenomena [37, 38]. Recently, however, their effectiveness has been remarked also beyond social simulation, particularly in domains where traditional techniques are typically adopted [39], such as parallel and distributed discrete-event system simulation, object-oriented simulation, and dynamic micro-simulation.

5.1   Agent-Based Models vs. Traditional Models

In general, simulations based on the agent paradigm integrate aspects that can be found both in *micro* and *macro* techniques for simulation.

On the one side, in the same way as micro techniques, agent-based approaches model specific behaviour of individual entities or components. This can be contrasted to macro simulation techniques which are typically based on traditional ordinary differential equation models where the characteristics of a *population* are averaged together, and the model attempts to simulate changes in the averaged characteristics of the whole population. Typically, in these approaches a whole population of entities is subdivided in sub-populations which are *homogeneous*, i.e. the members of each sub-population are indistinguishable from each other. By contrast, in agent-based models there is no such aggregation: the (spatially) distributed population is heterogeneous and consists of distinct agents, each with its own state and interaction behaviour, which typically evolves with time. Thus, in macro simulation the set of individuals is viewed as a structure that can be characterised by a number of variables, whereas in agent-based as well as micro simulations the structure is viewed as an emergent property, outcome of the interactions between the individuals.

As described in Parunak et al. [40] : "...agent-based modelling is most appropriate for domains characterised by a high degree of localization and distribution and dominated by discrete decision and local interaction. Equation-based modelling is most naturally applied in systems that can be modelled centrally, and in which the dynamics are dominated by physic laws rather that information processing...". Here, we promote a conceptual framework for simulating biological systems heavily based on the interaction of autonomous, localised, distributed components, for which the agent-based approach is then the ideal framework.

On the other side, in the same way as in macro techniques, agent-based approaches promote the investigation of systemic properties that cannot be understood at the individual component level, but require the introduction of new categories for their description. In other words, agent-based approaches make it possible to simulate and analyse *emergent* properties, which can be understood as properties of the ensemble of the components in the overall. Differently from traditional micro-techniques, which typically adopt primitive forms of interaction among the individuals with very simple behaviours, agent-based approaches allows the interaction among the components to be modelled at the *knowledge level*, as communications driven by the semantic of the information exchanged.

5.2 Modelling Biological Systems as MAS

The notion of agent has been described in several ways in literature [41, 42, 43], with different acceptation according to the research field where it has been considered—from distributed artificial intelligence (DAI) to software engineering and concurrent/distributed systems, from social/psychological/economic sciences to computer supported cooperative work (CSCW). Among the core properties of the agent abstraction, in this context we focus on *autonomy* and *interaction*.

As an autonomous entity, an agent encapsulates the execution of independent activities or tasks within its environment. In the software engineering context, autonomy is characterised by the encapsulation of behaviour control: as for the object abstraction, agents encapsulate a state and a behaviour; differently from objects, agents have full control of both (objects only of their state). So agents work by autonomously executing their tasks, concurrently to the work of the other agents.

As a situated entity, an agent is a persistent entity immersed within and interacting with an environment, which is typically open and dynamic. Interaction—in its most wide characterisation [44]—is a fundamental dimension of the agent paradigm: generally speaking, an agent interacts with its environment by means of actions and perceptions, which enable the agent to partially observe and control the environment. In the literature, further characterisations have been attributed to the agent abstraction: examples are pro-activity (as the capability of taking initiative) and social ability (the adoption of high level languages for inter-agent communication). Heterogeneous computational/behavioural models have led to different forms of agent classification: examples are *intelligent* agents—when the agent behaviour is defined in terms of high level cognitive/mentalistic structures and processes, with an explicit symbolic representation of knowledge, interaction and related reasoning processes—and *reactive* agents—typically characterised by sub-symbolic (such as neural networks) or imperative computational models.

So, how can the agent abstraction be exploited for modelling biological systems? As a first, obvious point, biological systems typically feature by a number of complex and concurrent activities. So, in biological system simulations agents can be suitably adopted for modelling at the appropriate level of abstractions such activities (tasks), or, even better, the biological components that are responsible for them. Agent abstraction makes it possible to model directly both reactive and proactive behaviour of biological components, accounting for both internal and interactive activities. The different kind of interaction within the biological environment can be suitably modelled in terms of actions and perceptions (including communication), representing the flow of chemical material and electric stimuli. The heterogeneity of the available agent models (e.g. cognitive, reactive,...) makes it possible to have different ways to describe the behaviour of biological components, according to the aspects that we are interested in considering.

Then, besides individual bio-components, a key-aspect for our simulations is modelling the environment where the bio-components are immersed, and, more generally, the biological *system* in the overall. By exploiting the MAS paradigm, such aspects can be modelled with first-class abstractions that have been introduced to model the agent environment and, in particular, the environmental structures or *artifacts* which mediate agent interaction. Agent interaction can be modelled both as direct communication among agents using some kind of Agent Communication Language (ACL), and as mediated interaction exploiting the mediating artifacts which are part of their environment [45]. A mediating artifact can range from a simple communication channel to a shared data-structure, from a shared blackboard to a scheduler, useful for agents to synchronise their tasks. The latter ones, in particular, are examples of *coordination artifacts*, i.e. mediating artifacts providing specific coordination functionalities [46].

Thus, MASs provide for the appropriate abstraction level to model a biological system in the overall: if an agent represents an individual component of the systems, the overall MAS including environmental abstractions captures the overall set of the biological components and their environment, also including the structures involved in their interaction. Mediating and coordination artifacts in particular can be adopted to model the various patterns of interaction that can be found in biological processes—such as the many examples that can be found in the KEGG Pathway database [47].

It is worth noting that the model of (mediating) artifacts can be crucial for creating a simulation where the overall emergent and stable behavior can be reproduced. For instance, works on complex system simulation [48, 49] describe a case study where the correct behaviour of the simulated system could be obtained only by properly modelling knowledge structures shared by agents, which here can be framed as sorts of mediating artifacts. In particular, the above-cited research points out the benefits of such structures in simulation environments in terms of the dramatic enhancement of the probability of reproducing stable and acyclic overall behaviours. Generalising this case, we expect to have analogous benefits in adopting first-class abstractions to model and control interaction in biological systems.

Finally, in the context of the MAS paradigm, high-level organisational models—based on concepts such as roles, inter-role relationships, groups and societies—are used to characterise the structure as well as the structure relationships among the components. The *agent society* notion can be used here to define an ensemble of agents *and* the environmental abstractions involved in the *social task* characterising the society: a social task accounts for the coordinated execution and interaction of agent individual tasks, toward the achievement of an overall (social) objective. The notion of agent society can be suitably adopted for scaling with complexity, identifying different levels of descriptions of the same system: what is described at one level as an individual agent, could be described at a more detailed level as a society of agents (zooming in)—so, as an ensemble of agents plus their environmental abstractions—and vice-versa (zooming out). These modelling features could be then exploited for the simulation of biological systems involving different description levels, each one characterised by different sorts of emerging phenomena.

5.3   Engineering Biological System Simulations

MAS paradigm can be effective not only for technologies to build simulations, but first of all for devising a *methodology* covering the whole simulation engineering spectrum, from design to development, execution and runtime (dynamic) control. Critical points of biological systems—concerning structures, activities, interactions—can be captured directly by abstractions that *are kept alive from design to execution time*, supported by suitable infrastructures [50].

The simulation can then be framed as an on-line experiment, where the scientist can observe and interact dynamically with the system and its environment, both by changing its structure—by introducing for instance new agents representing biological components or removing existing ones—and global biological processes—by acting on the mediating/coordination artifacts. In particular, the *control* of mediating artifacts at run-time is the key for supporting the analytical and synthetical processes, promoting system behavior analysis and system control as defined by Kitano in [1]. In this case, controlling the mediating artifacts implies the possibility of *(a)* inspecting the dynamic state of the artifact, and *(b)* adapting the artifact, by changing its state and—more radically—changing its mediating behavior. On the one side, inspecting the dynamics of system interactions is fundamental for supporting the analysis of system behavior, enabling the identification, observation and monitoring of (emerging) patterns of interaction. On the other side, the possibility of dynamically adapting the state and behaviour of the mediating artifacts can be useful, for instance, to observe system reaction—in term of stability and robustness—to unexpected events concerning the component interactions; but also to enact some forms of feedback, analogously to feedbacks in control system theory: coordination processes gluing the components can be adapted dynamically according to (unexpected) events occurring in the system or in its environment.

The capability of controlling mediating and coordination artifacts is meant to be provided both to humans (scientists and simulation engineers) and artificial cognitive agents—the latter typically featuring reasoning capabilities supporting automatic forms of system behaviour analysis.

In this overall scenario, MAS *infrastructures*, and agent coordination infrastructures in particular, play a fundamental role [50]. Agent infrastructures provide basic services to sustain the agent life-cycle, supporting dynamic agent spawning, death, (possibly) mobility and also some form of native communication services. Well-known examples are JADE [7], a FIPA-compliant platform, RETSINA [51] and Hermes [8]. Coordination infrastructures, such as TuCSoN [52], provide instead specific services to support agent interaction and coordination [50]. These services are typically concerned with the access and management of different kind of mediating/coordination artifacts, in order to make them shared and used concurrently by agents. In particular, they can provide support (tools, interfaces) for artifacts control [9].

5.4   Formal Specification and Verification of Biological Systems

As already discussed in the previous sections, the simulation of a complex system first of all requires the construction of the system model. Different modelling techniques are available and can be used, and their choice is usually determined by the peculiarities of the application domain. Once the model is defined, in order to prove that the corresponding system is correct we must resort to the use of formal methods to rigorously define and analyse functionalities and behaviour of the system itself. In the context of systems biology, formal modelling is becoming of paramount importance. In fact, biologists have begun themselves to define (semi-)formal notations [26, 4] mainly to guarantee the unambiguous sharing and integration of models developed by different users. More formally, computer scientists are working on abstract notations, useful to build and to validate biological models [53, 54, 18, 30, 10]. An exhaustive referenced list of formal methods for systems biology can be found in [10].

Agent-oriented organisation proves to be natural and useful in representing three different views (models) of a biological system: functional, dynamic and static-structural [11]. There have been a few attempts to define formal methods to prove that once a MAS has been defined, it satisfies peculiar properties of the modelled system, it behaves correctly for the purpose it has been created, and in presence of noise it can also predict the system behaviour [55, 56]. We consider formal methods as being essential tools also for any agent-based conceptual framework, which should include among its purposes the answer to fundamental questions related to system functionalities like, for instance: who are the components (actors) involved in such system functions? Which is the role played by such components? How would such components behave in presence / absence of other given components? How do they participate to a given function, with which other agents do they interacts under certain circumstance, or along their whole life? Furthermore, for static-structural analysis, we would like to be able to know which type of variables describe (or what elements are in) the environment in a given situation, which dependency or relationship does exist between a set of variables, which type of variable is inferred as result of a certain function, and whether is it feasible that new type could be taken into account. Preliminary results on integrated approach are described in [22, 11].

6   A Glance into a Biological Example

As we already emphasised, a so-deeply complex system like a cell can be described and simulated at different abstraction levels. The agent paradigm is expressing all its power and elegance when founding the modelling at the molecular level ("molecules-as-agents"). In our vision, every basic component of the cell should be an agent. A population of molecular agents is living in a virtual cell, being allowed to walk randomly through the cellular simulated spaces and to communicate with all the other cellular agents they encounter. For example, in metabolic pathways when an enzyme-agent finds all its necessary counterparts, then the chemical reaction it is able to execute can take place, and all the participating agents undergo some planned transformation. From a theoretical point of view, this sounds more or less feasible. Some successful attempts have been already made: for instance, Pogson et al. [57] describe the implementation and the evolution of a population of about five thousands receptors and sixty thousand NF-kB molecules, randomly moving and interacting in a spatial context. However, the limits of current technology (difficulty to manage thousands of agent species, huge numbers of agent instances and especially communications and interactions) suggest that, at least for the close future, the cellular process should be examined at a higher abstraction level.

To illustrate the approach promoted by the framework, we consider as a case study carbohydrate oxidation, detailed described in [11]. The carbohydrate oxidation is the process by which a cell produces energy through chemical transformation of carbohydrates [58].

The carbohydrate oxidation pathway is performed by two main cell components: cytoplasm and mitochondria. Cytoplasm is the viscid, semifluid component between the cell membrane and the nuclear envelope: it is responsible for the first stage of carbohydrate molecule transformation. This process, called glycolysis, converts molecules of *glucose* in those of *pyruvate*. Then, pyruvate, under anaerobic conditions, is reduced either by lactic fermentation to *lactate*, or by alcoholic fermentation to *ethanol*, whereas in the aerobic respiration pyruvate is reduced to water and carbon dioxide.

A mitochondrion consists of two main sub-components, the inner mitochondrial membrane (IMM) and the mitochondrial matrix (MM). It is responsible for the aerobic respiration process, which consists in transporting the pyruvate produced during the glycolysis through the IMM into the MM, where the partial oxidation process degrades the pyruvate, and combines it with *coenzyme A* to form *acetyl coenzyme A*. Finally, the IMM is responsible for the Kreb's cycle that forms part of the break-down of carbohydrates, fats and proteins into carbon dioxide and water, in order to generate energy.

Metabolic Pathways as Social/Environmental Activities. Following the agent-oriented conceptual framework proposed in this article, the cell can naturally be modelled as a society of autonomous, interacting agents: the cytoplasm and the mithocondria, each of which represented by IMM and MM. Every agent behaves in active way, by simulating the functions of the biological component it represents, and in pro-active way by perceiving and interacting with the environment where it is situated.

Figure 1 describes, by means of a UML stereotype, the cell functions involved in the oxidation of carbohydrate: glycolysis, alcoholic fermentation etc. as a society of simulation agents interacting with the environment. Each agent (identified by a transparent ovals shape) can play more than one roles, e.g. one for each functions of the associated component. In the example, the cytoplasm is an agent responsible for the glycolysis, the lactic and alcoholic fermentations for which it plays three different roles. The cytoplasm, at different, more fine-grained level of abstraction, can be seen (zooming-in) as an ensemble of agents plus its interaction environment, and so on until the basic computational unit is reached. The glycolysis is a chemical reactions that transforms environment elements, i.e. *glucose* molecules into pyruvate molecules. This transformation can be intuitively modelled as a natural law of the environment in terms of its chemical reactions:

glucose+ 2atp+ 4adp+ 2Pi + 2NADox$\rightarrow$ 2pyruvate+ 2adp+ 4atp+ 2NADrid+ 2H$_2$O

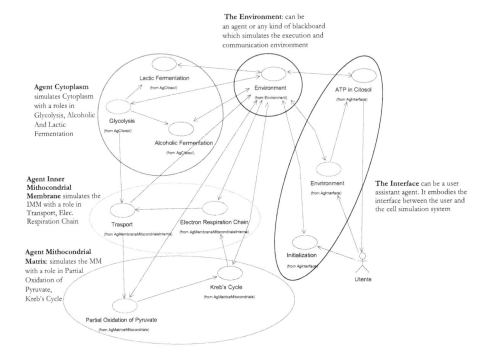

Fig. 1. Agent society of the carbohydrate oxidation cell's functions

When the rule is triggered, it automatically produce and consume from the environment molecules of glucose, pyruvate, atp, adp, ... (e.g. messages in a blackboard, or tuples in a tuple space).

So, in order to give a simple and intuitive example, we may for instance exploit the TuCSoN infrastructure for agent coordination [52], and use its ReSpecT tuple centres [59] to enforce rules for chemical reactions as agent coordination rules. Then a tuple of the form molecule(X) would represent the different types of elements, e.g. tuples of the type: .plus.1em minus.1em

molecule(glucose) molecule(atp) molecule(adp) molecule(pyruvate) ...

The production and consumption of molecules in the environment would then correspond to put or get tuples from the tuple centre. Thus, a chemical reaction could be implemented as a ReSpecT logic rule through defining the functional behaviour of the corresponding agent. So, the glycolysis could be implemented by the logic rule *"if in the environment there is one molecule of* glucose, *two of* atp, *four of* adp *and ... then the glycolysis can be activated"*, to which the cytoplasm reacts, any time a new *molecule(X)* is introduced in its environment, by the following behaviour, implemented as the ReSpecT specification sketched below:

```
reaction(out(molecule(X)),(
  in_r(molecule(glucose)),
  in_r(molecule(atp)),
```

```
in_r(molecule(atp)),
...
out_r(molecule(pyruvate)),
out_r(molecule(pyruvate)),
out_r(molecule(adp)),
out_r(molecule(adp)),
...
)).
```

Of course, once the simulation model for the oxidation of carbohydrate has been designed and implemented, it is important to know whether the structure and function of the model accurately reflects the biological system.

7   Conclusion and Future Directions

In this paper, we have investigated on the application of the agent-oriented paradigm to define a conceptual framework for engineering simulations of biological systems. We have discussed how a biological system can be seen as a society of autonomous interactive components, whose complexity depends on the level of description. Then, we have introduced the notion of agent society as suitable for identifying different levels of abstraction, and for scaling with the system complexity. As a future work, based on the proposed approach, we aim at defining a complete methodology covering the whole process of simulation engineering, from design to development, testing, and validation of both behavioural and static-structural models of a biological system.

References

[1] Kitano, H.: Foundations of Systems Biology. MIT Press (2002)
[2] Finkelstein, A., Hetherington, J., Li, L., Margominski, O., Saffrey, P., Seymour, R., Warner, A.: Computational challenges of systems biology. IEEE Computer **37** (2004) 26–33
[3] Systems Biology Workbench. (http://sbw.sourceforge.net)
[4] Systems Biology Modelling Language. (http://www.sbml.org)
[5] Hucka, M., Finney, A., Sauro, H., et al.: The systems biology markup language (SBML): a medium for representation and exchange of biomedical network models. Bioinformatics **19** (2003) 524–531
[6] CellML. (http://www.cellml.org)
[7] Bellifemine, F., Poggi, A., Rimassa, G.: Developing multi-agent systems with a fipa-compliant agent framework. Software Practice and Experience **31** (2001) 103–128
[8] Corradini, F., Merelli, E.: Hermes: Agent-based middleware for mobile computing. In Bernardo, M., Bogliolo, A., eds.: Formal Methods for Mobile Computing. Volume 3465 of LNCS. Springer (2005) 234–270 5th International School on Formal Methods for the Design of Computer, Communication, and Software Systems, SFM-Moby 2005, Bertinoro, Italy, April 26-30, 2005, Advanced Lectures.
[9] Denti, E., Omicini, A., Ricci, A.: Coordination tools for the development of agent-based systems. Applied Artificial Intelligence **16** (2002)

[10] Cardelli, L.: Abstract machines of systems biology. LNCS Transactions on Computational Systems Biology (2005) In this volume.
[11] Corradini, F., Merelli, E., Vita, M.: A multi-agent system for modelling the oxidation of carbohydrate cellular process. In Gervasi, O., et al., eds.: Computational Science and Its Applications – ICCSA 2005. Volume 3481 of LNCS., Springer (2005) 1265–1273 International Conference, Singapore, May 9-12, 2005, Proceedings, Part II.
[12] d'Inverno, M., Saunders, R.: Agent-based modelling of stem cell organisation in a niche. In Brueckner, S.A., Di Marzo Serugendo, G., Karageorgos, A., et al., eds.: Engineering Self-Organising Systems: Methodologies and Applications. Volume 3464 of LNAI. (2005)
[13] Walker, D., Southgate, J., Hill, G., Holcombe, M., Hose, D.R., Wood, S.M., Mac Neil, S., Smallwood, R.H.: The epitheliome: agent-based modelling of the social behaviour of cells. Biosystems **76** (2004) 89–100
[14] Corradini, F., Mariani, L., Merelli, E.: An agent-based approach to tool integration. Journal of Software Tools Technology Transfer **6** (2004) 231–244
[15] Merelli, E., Young, M.: Validating MAS models with mutation. In: 1st International Workshop on Multi-Agent Systems for Medicine, Computational Biology, and Bioinformatics (MAS*BIOMED'05), AAMAS'05, Utrecht, The Netherlands, July 25 (2005)
[16] Kacprzak, M., Lomuscio, A., Penczek, W.: Verification of multiagent systems via unbounded model checking. In Jennings, N.R., Sierra, C., Sonenberg, L., Tambe, M., eds.: 3rd international Joint Conference on Autonomous Agents and Multiagent Systems (AAMAS 2004). Volume 2., New York, USA, ACM (2004) 638–645
[17] De Jong, H.: Modeling and simulation of genetic regulatory systems: a literature review. Journal of Computational Biology **9** (2002) 67–103
[18] Antoniotti, M., Mishra, B., Piazza, C., Policriti, A., Simeoni, M.: Modeling cellular behavior with hybrid automata: Bisimulation and collapsing. In Priami, C., ed.: Computational Methods in Systems Biology. Volume 2602 of LNCS. Springer (2003) 57–74 1st International Workshop, CMSB 2003, Rovereto, Italy, February 24-26, 2003. Proceedings.
[19] Mendes, P.: GEPASI: a software package for modelling the dynamics, steady states and control of biochemical and other systems. Computer Applications in the Biosciences **9** (1993) 563–571
[20] Tomita, M., Hashimoto, K., Takahashi, K., Shimizu, T.S., Matsuzaki, Y., Miyoshi, F., Saito, K., Tanida, S., Yugi, K., Venter, J.C., Hutchison, C.A.: E-CELL: software environment for whole-cell simulation. Bioinformatics **15** (1999) 72–84
[21] Antoniotti, M., Policriti, A., Ugel, N., Mishra, B.: S-systems: extended s-systems and algebraic differential automata for modeling cellular behavior. In Sahni, S., Prasanna, V., Shukla, U., eds.: High Performance Computing – HiPC 2002. Volume 2552 of LNCS. Springer (2002) 431–442 9th International Conference, Bangalore, India, December 18-21, 2002. Proceedings.
[22] Peleg, M., Yeh, I., Altman, R.: Modeling biological process using workflow and Petri nets. Bioinfomatics **18** (2002) 825–837
[23] Matsuno, H., Tanaka, Y., Aoshima, H., Doi, A., Matsui, M., Miyano, S.: Biopathways representation and simulation on hybrid functional Petri net. Silico Biology **2003** (3) 389–404
[24] Koch, I., Junker, B.H., Heiner, M.: Application of Petri net theory for modelling and validation of the sucrose breakdown pathway in the patato tuber. Bioinformatics **21** (2005) 1219–1226

[25] Nagasaki, M., Doi, A., Matsuno, H., Miyano, S.: A versatile Petri net based architecture for modeling and simulation of complex biological processes. Genome Informatics **16** (2004)

[26] Nagasaki, M., Onami, S.: Bio-calculus: Its concept, and an application for molecular interaction. Genome Informatics **10** (1999) 133–143

[27] Regev, A., Silverman, W., Shapiro, E.: Representation and simulation of biochemical processes using the pi-calculus process algebra. In: Pacific Symposium of Biocomputing 6 (PSB 2001). (2001) 459–470

[28] Regev, A.: Representation and simulation of molecular pathways in the stochastic pi-calculus. In: 2nd workshop on Computation of Biochemical Pathways and Genetic Networks, Heidelberg, Germany (2001)

[29] Chiarugi, D., Curti, M., Degano, P., Marangoni, R.: VICE: a VIrtual CEll. In Danos, V., Schachter, V., eds.: Computational Methods in Systems Biology. Volume 3082 of LNCS., Springer (2004) 207–220 International Conference, CMSB 2004, Paris, France, May 26-28, 2004, Revised Selected Papers.

[30] Cardelli, L.: Brane calculi: Interactions of biological membranes. In Danos, V., Schachter, V., eds.: Computational Methods in Systems Biology. Volume 3082 of LNCS., Springer (2004) 257–278 International Conference, CMSB 2004, Paris, France, May 26-28, 2004, Revised Selected Papers.

[31] Milner, R.: Communication and Concurrency. Prentice Hall (1989)

[32] Gonzalez, P., Cardenas, M., Camacho, D., Franyuti, A., Rosas, O., Otero, J.: Cellulat: an agent-based intracellular signalling model. BioSystems **68** (2003) 171–185

[33] Lazebnik, Y.: Can a biologist fix a radio?—Or, what I learned while studying apoptosis. Cancer Cell **2** (2002) 179–182

[34] Regev, A., Shapiro, E.: Cellular abstractions: Cells as computation. Nature **419** (2002) 343

[35] Zambonelli, F., Omicini, A.: Challenges and research directions in agent-oriented software engineering. Autonomous Agents and Multi-Agent Systems **9** (2004) 253–283

[36] Jennings, N.: An agent-based approach for building complex software systems. Communication of ACM **44** (2001) 35–41

[37] Conte, R., Edmonds, B., Moss, S., Sawyer, K.: Sociology and social theory in agent based social simulation: A symposium. Computational and Mathematical Organizational Theory **7** (2001)

[38] Gilbert, N., Conte, R., eds.: Artificial Societies: the computer simulation of social life. UCL Press (1995)

[39] Davidsson, P.: Agent based social simulation: A computer science view. Journal of Artificial Societies and Social Simulation **5** (2002)

[40] Parunak, V.D., Savit, R., Riolo, R.L.: Agent-based modelling vs. equation based modelling: A case study and users' guide. In Sichman, J.S., Gilbert, N., Corte, R., eds.: Multi-Agent Systems and Agent-Based Simulation. Springer-Verlag (1998) 10–26

[41] Jennings, N., Wooldridge, M., eds.: Agent Technololgy: Foundations, Applications, and Markets. Springer-Verlag (1998)

[42] Wooldridge, M.J., Jennings, N.R.: Intelligent agents: Theory and practice. The Knowledge Engineering Review **10** (1995) 115–152

[43] Jennings, N.R.: On agent based software engineering. Artificial Intelligence **117** (2000) 277–296

[44] Wegner, P.: Coordination as constrained interaction. In Ciancarini, P., Hankin, C., eds.: Coordination Languages and Models. Volume 1061 of LNCS. Springer-Verlag (1996) 305–320 1st International Conference (COORDINATION'96), Cesena, Italy, April 15-17, 1996.

[45] Ricci, A., Viroli, M., Omicini, A.: Environment-based coordination through coordination artifacts. In Weyns, D., Parunak, H.V.D., Michel, F., eds.: Environments for MultiAgent Systems. Volume 3374 of LNAI. Springer-Verlag (2005) 190–214 1st International Workshop (E4MAS 2004), New York, NY, USA, 19 July 2004. Revised Selected Papers.

[46] Omicini, A., Ricci, A., Viroli, M., Castelfranchi, C., Tummolini, L.: Coordination artifacts: Environment-based coordination for intelligent agents. In Jennings, N.R., Sierra, C., Sonenberg, L., Tambe, M., eds.: 3rd international Joint Conference on Autonomous Agents and Multiagent Systems (AAMAS 2004). Volume 1., New York, USA, ACM (2004) 286–293

[47] KEGG: PATHWAY database. (http://www.genome.ad.jp/kegg/pathway.html)

[48] Cliff, J., Rocha, L.M.: Towards semiotic agent-based models of socio-technical organizations. In Sarjoughian, H., et al., eds.: AI, Simulation and Planning in High Autonomy Systems (AIS 2000), Tucson, AZ, USA (2000) 70–79

[49] Richards, D., Richards, W.A., McKey, B.: Collective choices and mutual knowledge structures. Advances in Complex Systems **1** (1998) 221–236

[50] Omicini, A., Ossowski, S., Ricci, A.: Coordination infrastructures in the engineering of multiagent systems. In Bergenti, F., Gleizes, M.P., Zambonelli, F., eds.: Methodologies and Software Engineering for Agent Systems: The Agent-Oriented Software Engineering Handbook. Kluwer Academic Publishers (2004) 273–296

[51] Sycara, K., Paolucci, M., van Velsen, M., Giampapa, J.: The RETSINA MAS infrastructure. Autonomous Agents and Multi-Agent Systems **7** (2003) 29–48

[52] Omicini, A., Zambonelli, F.: Coordination for Internet application development. Autonomous Agents and Multi-Agent Systems **2** (1999) 251–269 Special Issue: Coordination Mechanisms for Web Agents.

[53] Danos, V., Krivine, J.: Formal molecular biology. Theoretical Computer Science **325** (2004) 69–110

[54] Regev, A., Panina, E.M., Silverman, W., Cardelli, L., Shapiro, E.: Bioambients: An abstraction for biological compartments. Theoretical Computer Science. Special Issue on Computational Methods in Systems Biology. Elsevier **325** (2004) 141–167

[55] d'Inverno, M., Luck, M.: Understanding Agent Systems, 2nd Ed. Springer (2004)

[56] Odell, J., Parunak, H., Bauer, B.: Extending uml for agents. In: Agent-Oriented Information Systems Workshop at the 17th National conference on Artificial Intelligence. (2000)

[57] Pogson, M., Holcomb, M., Qwarnstrom, E.: An agent based model of the NF-kB signalling pathways. In: 5th International Conference on Systems Biology (ICSB2004), Heidelberg, Germany (2004)

[58] Garret, R., Grisham, C.: Biochemistry. Sunder College Publishing (1995)

[59] Omicini, A., Denti, E.: From tuple spaces to tuple centres. Science of Computer Programming **41** (2001) 277–294

Genetic Linkage Analysis
Algorithms and Their Implementation

Anna Ingolfsdottir[1,2,3] and Daniel Gudbjartsson[1]

[1] DeCode Genetics, Reykjavik, Iceland
[2] Deparment of Engineering, University of Iceland, Reykjvk, Iceland
[3] **BRICS** (**B**asic **R**esearch **i**n **C**omputer **S**cience),
Centre of the Danish National Research Foundation,
Department of Computer Science, Aalborg University, Aalborg, Denmark
`annaing@hi.is`

Abstract. *Linkage analysis* is a well established method for studying the relationship between the pattern of the occurrence of a given biological trait such as a disease and the inheritance pattern of certain genes in a given family. In this paper we give some improved algorithms for the genetic linkage analysis. In particular we offer an MTBDD based version of these algorithms that turns out to be a considerable improvement of the existing ones both in terms of time and space efficiency. The paper also contains a formal mathematical or computational description of the linkage analysis that gives a deeper insight into the area and therefore a better understanding of the related problems and how to solve them.

1   Introduction

Genetic analysis of human traits is substantially different from the analysis of experimental plant and animal traits in that the observed pedigrees must be taken as given, but are not controlled as for the experimental organisms. This creates the need for methodologies such as *linkage analysis*, which is a well established method for studying how given biological traits cosegregate with the inheritance of certain parts of the genome in given families. In order to explain the main idea of the method we give a short survey of the biological model it is based on.

Linkage Analysis: The genetic material, i.e. the *genome*, of an organism is represented in the DNA (deoxyribonucleic acid) which is a chain of simpler molecules, *bases*. The genetic material is organized in the organism in such a way that each cell has a few very long DNA molecules, called *chromosomes*. Human beings are diploids, meaning that every individual carries two almost identical copies of each chromosome. According to Mendel's law, at each position or *locus*, there are equal probabilities of the genetic material of each of the two chromosomes of a parent being transmitted to a child. Before the parents pass their chromosomes on to their offspring, the chromosomes mix in a process called *meiosis*. The mixing involves large chunks of each chromosomes being transmitted intact, the transmission only being interrupted by transmissions of large chunks of the other chromosome. Where the transmission moves between the two

C. Priami et al. (Eds.): Trans. on Comput. Syst. Biol. III, LNBI 3737, pp. 123–144, 2005.
© Springer-Verlag Berlin Heidelberg 2005

chromosomes a *crossover* is said to have occurred. The genetic distance between two loci is defined as the expected number of crossovers that will occur in a single meiosis between the two loci and is given in the unit *Morgan*. The autosomal human chromosomes are between .7 and 2.7 Morgans in length. If an odd number of crossovers occurs between two loci, then genetic material from different chromosomes is transmitted and a *recombination* is said to take place. The *recombination fraction* between two loci is defined as the probability of a recombination occurring between them. It is commonly assumed that crossovers follow a Poisson process, in which case a simple mapping between genetic distance and recombination fractions exists, called the Haldane mapping. Two loci are said to be *linked* if the recombination fraction between them is less than $\frac{1}{2}$. It is worth noting that loci on different chromosomes are unlinked. For a more detailed description of the field of linkage analysis see for instance [16].

The dependence between the transmissions statuses of loci that are physically close in the genome is the basis of linkage analysis. The aim of linkage analysis is to identify loci where the *inheritance pattern* deviates substantially from Mendelian inheritance. The dependence described above allows us to study accurately the transmissions of the whole genome by investigating only a small number of loci for which information is available.

Linkage analysis is performed on a *pedigree*. The set of members V of each pedigree is divided into the set of founders (F), where neither of the parents belongs to the pedigree, and non-founders (N) where both parents do. We may assume that either both or neither of the individual's parents belong to the pedigree as we can always add the missing one. When linkage analysis is performed, a number of known loci or *markers* are chosen. More precisely, a marker is a locus where the set of possible states, or *alleles*, and the frequency of each of these alleles, are known in the population. The recombination fraction between these markers must be known too. Such a set of markers with the information described above is referred to as a *marker map*. Next those individuals of the pedigree that are accessible are *genotyped* for each of these markers. This results in a sequence of allele pairs, one for each marker under investigation. As we cannot directly observe from which parent each of the allele comes, i.e. the *phase* of the alleles is not observable, these pairs are unordered. Because the phase of the genotype is in general not known and because not all individuals in a pedigree can be genotyped, the precise inheritance pattern, i.e. which alleles are inherited from which parents, is represented by a probability distribution.

If only one marker is investigated at the time, independently of other markers, we talk about *single point analysis*. This amounts to calculating the probability of a given inheritance pattern at a given marker, given the genotypes at that particular marker only. The genotypes of the markers do in general not give complete information about transmissions. Therefore, to get as precise probability distributions over the inheritance patterns as possible, the information about the genotype at the remaining markers is also taken into account and we calculate the probability of a given inheritance pattern at a given marker given the genotypes for all markers. In this case we talk about *multi point analysis*. The multi point probabilities are derived from the single point ones by taking the probability of recombination into account.

Related Work: Most contemporary methods for doing multi point linkage analysis are based either on the Elston-Stewart algorithm or the Lander-Green algorithm. The Elston-Steward algorithm was introduced in [8] and has running time that grows linearly with the number of individuals in the pedigree but exponentially in the number of markers under investigation. The algorithm is based on parametric linkage analysis. VITESSE is a state of the art software package based on this algorithm and, with the most recent improvements, it can handle approximately 6-9 markers in a moderately large pedigree [22].

The second algorithm, suggested by Lander and Green in [17], is the one we focus on in this study. It is based on Hidden Markov Models (HMMs [23]), a generalization of the standard Markov Models (MMs) where the states are "hidden". This means that in the HMMs, unlike in the MMs, the states are not explicitly observable and instead each state is represented as a probability distribution over possible observations given that state. In our case the time corresponds to the position of the markers (that are ordered from left to right), states are modes of inheritance, or *inheritance patterns*, expressed as bit vectors at each marker and the observations are sequences of genotypes given at each marker. With this formulation the standard theory of HMMs can be directly applied and implemented.

The Lander-Green algorithm supports non-parametric linkage analysis (NPL).

The algorithm has two main phases, the single point phase and the multi point one. In the single point phase, for a given marker or locus, the probability of each inheritance pattern is calculated, given only the genotypes for this particular marker. This is obtained by calculating the set of all allele assignments for the whole pedigree that are compatible with the given sets of genotypes and the given inheritance pattern. From this the probability is calculated from the frequency distribution over the alleles in the population.

The multi point probabilities are obtained by an iterative process where the probability distribution at each marker is multiplied with its *left* and *right contribution*. The left contribution is obtained, like for standard Markov Chains, by taking the convolution of the distribution at the next marker to left and the transition probabilities.

The original version of the Lander-Green algorithm had the running time $O(m2^{4n})$, where m is the number of markers and n is the number of non-founders in the pedigree and was implemented in the software package MAPMAKER. Since then several improvements have been made on this algorithm towards more efficient use of time and space: By applying a divide-and-conquer technique, Kruglyak et. al. provided a version that runs in $O(mn^2 2^{n-2})$. In [13] Idury and Elston put forward a version where they explore the regularity of the transition matrix by writing it as a Kronecker product of simple basic matrices. This version of the algorithm runs in time $mn2^n$. Kruglyak et. al. [14] improved this result by recognizing that in ungenotyped founders, there is no way of distinguishing between the maternal and paternal genetic material. Thus inheritance patterns that differ only by phase changes in the founders are completely equivalent in the sense that they have the same probability independent of the genotypes. By treating such an equivalence class as one pattern, i.e. using *founder reduction*, the time (and space) complexity reduces by a factor of 2^f where f is the number of founders in the pedigree. This version of the algorithm was incorporated in the software package GENHUNTER.

In [15] Kruglyak and Lander suggested to use Fast Fourier Transform (FFT) for additive groups to reduce the complexity. By incorporating the founder reduction into this implementation they obtained a version of the algorithm with the the same time complexity as the Idury-Elston version with founder reduction.

In [10] Gudbjartsson et.al. further improved on the performance of the Lander-Green algorithm by not distinguishing between the individuals of a founder couple and thereby reduce the complexity further by 2^c, where c is the number of such couples in the pedigree. Furthermore they implemented the single point calculation in a top down manner such that inconsistencies are detected as soon as they occur. Their approach is based on the FTTs and is implemented in the software package ALLEGRO.

The linkage software MERLIN [2] is based on the Idury-Elston algorithm but the inheritance vectors are represented as sparse binary trees. The trees have three kinds of nodes: standard nodes, symmetric nodes, that represent the case when both branches lead to identical subtrees and premature leaf nodes, indicating that all subsequent branches lead to exactly the same value. In this way some of the redundancies in the representations of identical substructures are avoided. This version of the algorithm includes founder reduction but not founder couple reduction.

The most recent advances of the Lander-Green algorithm enable the analysis of pedigrees of approximately 20-30 bits [19,2].

Our Contribution: In present paper we suggest some further improvements of the Lander-Green algorithm. It is a further development of the Idury-Elston version described above and uses a compact representation along the lines of the MERLIN algorithm but incorporate both founder and founder couple reductions. In our approach we take seriously the idea of minimizing the redundancy in the representation. Therefore, instead of using sparse binary trees we use Multi Terminal Binary Decisions Diagrams (MTBDDs) [6] which are a generalization of the more standard Binary Decision Diagrams (BDDs)[5], a formalism widely used in, for instance, software verification to deal with reduction of large state spaces. A BDD is a symbolic representation of boolean function by means of a directed acyclic graph where all identical substructures are shared. For the implementation we use a the standard CUDD software of [21].

When we calculate the left and right contribution, instead of performing matrix multiplication directly like in previous approaches, all calculations are done in terms of the CUDD operations which ensure optimal use of time and space. Furthermore, as we know that we are going to multiply the left (and right) contribution with the present distribution at the marker, we only need to calculate the results for the elements that have non-zero probability at that marker.

Another gain with our approach is that also when there is little information about the genotypes, the resulting MTBDDs become small as most of the inheritance patterns have the same probability and are therefore capture by the sharing of the structure.

Our implementation turns out to improve greatly on the existing ones both in terms of time and space efficiency and if the data is of suitable quality, i.e. the individuals that are genotyped are all genotyped on the same markers, as much as twice as big families can be analyzed using this tool compared to other tools. This greatly improves the power of the method of genetic linkage analysis. As a part of the development we provide a formal mathematical or computational model of pedigrees and genetic inheritance in those. This model serves as a base for the MTBDD encoding of the multipoint calculations and allows us to investigate the founder and founder couple reductions and include them into the algorithm. Furthermore the model we offer has given us a deeper understanding of the problems related to linkage analysis and should be useful in future investigation of the subject.

2   Pedigrees and Related Structures

In this section we introduce a few concepts we need to express our algorithms.

Pedigrees: We define a *pedigree* as a tree-like structure where the nodes represent the *members* of the pedigree and the arcs represent the parental relationships. As each individual has exactly one mother and one father, these relationships are represented by a function, the *parental* function, that takes inputs of the form $(\mathbf{n}, b)$, where $\mathbf{n}$ is a member (nonfounder) of the pedigree and b has either the value p or m; the function returns the father of $\mathbf{n}$ if $b = p$ and the mother if $b = m$. As explained in the introduction, each member of the pedigree is assumed to be either a *founder* of the pedigree, where neither of his parents belongs to the pedigree, or a *non-founder* where both parents belong to it. In the definition to follow, if $F : A \times B \longrightarrow C$ we write F_y for the function $x \mapsto F(x, y)$. Furthermore, following standard practice, for $f : A \longrightarrow B$ and $A' \subseteq A$ we write $f(A')$ for $\{f(a)|a \in A'\}$.

Formally a *pedigree* consists of a 3-tuple $P = \langle V, F, \phi \rangle$ where the following holds:

- V is a finite set of *members* (ranged over by $\mathbf{m}$) and $F \subseteq V$ is a set of *founders* (ranged over by $\mathbf{f}$) of the pedigree.
- $\phi : (V \setminus F) \times \{0, 1\} \longrightarrow V$ is the *parental function* where
  - nobody can be both a mother and a father: $\phi_p(V \setminus F) \cap \phi_m(V \setminus F) = \emptyset$ and
  - a member of the family is never its own ancestor: $\forall \mathbf{m} \in V.(\mathbf{m}, \mathbf{m}) \notin (\phi_p \cup \phi_m)^+$, where $+$ is the transitive closure operator on relations .

The set $N = V \setminus F$ is referred to as the set of *non-founders* of the pedigree (ranged over by $\mathbf{n}$). Figure 1 shows an example of a pedigree with two founders and two nonfounders. In what follows we assume a fixed pedigree $P = \langle V, F, \phi \rangle$ (unless stated otherwise).

Chromosomes: The genetic material of an individual is organized in several separate DNA strings which are called *chromosomes*. They come in pairs of almost identical copies (apart from the sex chromosomes, which are not considered here), one the individual inherits from his mother and one he inherits from his father. As described in the introduction, each of these chromosome is the result of a recombination between the chromosomes of the corresponding chromosome pair of the parent it is inherited from. Thus some parts of this chromosome comes from the individual's grandmother and others from his grandfather. This fact is the base for the methodology behind the genetic linkage analysis as will be described below.

Alleles: At each locus of the chromosome pairs, the genetic material appears in slightly different variations, which we refer to as *alleles*. We let $\mathbf{A}$ denote the set of alleles that can occur (given a fixed locus).

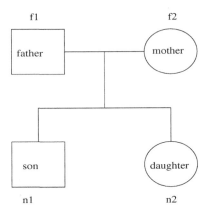

Fig. 1. A pedigree

3   The Single Point Case

In this section we describe how we calculate the single point probabilities for inheritance patterns from information about genotypes of some of the members of the pedigree.

Allelic Position: In linkage analysis we need to be able to express how the genetic material is passed on from parent to offspring (i.e. if the maternal allele at this locus of the chromosome is the mothers paternal or maternal one) without being explicit about the specific value of the allele. Therefore, at a given locus, we assume that the genetic material of the member $\mathbf{m}$ is placed at its *paternal allelic position* $(\mathbf{m}, p)$ and its *maternal allelic position* $(\mathbf{m}, m)$ (or just his positions for short); we define (for a fixed locus) the *allelic positions* (or positions) of the pedigree P as $\mathsf{Pos} = V \times \{p, m\}$.

Inheritance Patterns and Bits: An *inheritance pattern* is a complete specification of how each non founder inherits his genetic material from his parents, i.e. it indicates whether he inherits each of his alleles from the paternal or maternal positions of each of his parents. Thus, an *inheritance pattern* i is a function i : $\mathsf{Pos}_N \to \mathbb{B}$, where $\mathsf{Pos}_N = N \times \{p, m\}$ is the set of non-founder positions: $i(\mathbf{m}, m) = 0$ if, under meioses, individual $\mathbf{m}$ inherits his mother's paternal allele and similarly for the other combinations. For a given locus we refer to $i(\mathbf{n}, p)$ as the paternal bit of $\mathbf{n}$ and $i(\mathbf{n}, m)$ as his maternal bit.

Founder Source Positions: For a given member of the pedigree $\mathbf{n}$ and a given inheritance pattern i we can trace which pair of allelic positions of the founders the alleles of $\mathbf{n}$ are inherited from. We do this by defining the mapping $\mathsf{Source}^i : \mathsf{Pos} \to \mathsf{Pos}_F$ as follows:

$$\mathsf{Source}^i(\mathbf{n}, b) = \begin{cases} (\mathbf{n}, b) & \text{if } \mathbf{n} \in F \\ \mathsf{Source}^i(\phi(\mathbf{n}, b), \mathbf{b}(i(\mathbf{n}, b))) & \text{otherwise}, \end{cases}$$

where $\mathbf{b}(0) = m$ and $\mathbf{b}(1) = p$.

We write $\mathsf{Source}^i(\mathbf{n})$ as a shorthand for $\{\mathsf{Source}^i(\mathbf{n}, m), \mathsf{Source}^i(\mathbf{n}, p)\}$.

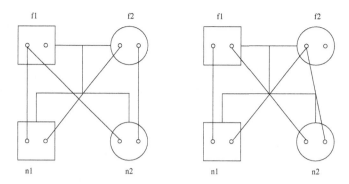

Fig. 2. Two different inheritance patterns for the same pedigree

Allele Assignment: An *allele assignment* is a function that maps each position in Pos to an allele: $\mathcal{A}$: Pos $\rightarrow$ **A**. A *founder allele assignment* is a function that maps each founder position to an allele: $\mathcal{F}$: $\text{Pos}_F \rightarrow$ **A**.

Remark 1. We note that if an allele assignment is to be compatible with a given inheritance pattern, then a given allelic positions must have the same allele assigned to it as the position it is inherited from (i.e. if, for instance, $i(\mathbf{m}, p) = 0$ then the position $(\mathbf{m}, p)$ and the position $(\mathbf{m}', m)$, where $\mathbf{m}'$ is the mother of $\mathbf{m}$, must be assigned the same allele).

If we fix the order of the set Pos, the inheritance pattern can be represented as a bit vector, an *inheritance vector*. Therefore we can freely choose between the notion of inheritance pattern and inheritance vector, depending on whether the order of the bits is fixed or not. Usually we let i, i_1, i' etc. denote inheritance patterns and u, v, w etc. inheritance vectors.

Figure 2 shows two different inheritance pattern for the pedigree in Figure 1. The left most dot in each circle/square indicates the paternal position of the member and the right most one the maternal position. If we fix the order of the non founder positions as $(\mathbf{n1}, p)(\mathbf{n1}, m)(\mathbf{n2}, p)(\mathbf{n2}, m)$, the inheritance pattern given in the left most figure is represented by 0001 and the right most one by 0010.

Genotypes: A *genotype* over the set of alleles **A** is defined as the subset of **A** that contains one or two elements. The set of genotypes over **A** is denoted by $\text{Geno}(\mathbf{A})$. *Genotype information* is a partial function $\mathcal{G}$: $V \rightharpoonup \text{Geno}(\mathbf{A})$, that associates a genotype to (some of the) members of the pedigree. The domain, $dom(\mathcal{G})$, of the function is referred to as the *set of genotyped members* of the pedigree. If $\mathbf{n} \in dom(\mathcal{G})$ we say that v is genotyped (at that marker).

Genotype Assignment: The set of genotypes over **A**, $\text{Geno}(\mathbf{A})$, is defined as the family of subsets of **A** that contain no more than two elements. Genotype assignment is a partial function $\mathcal{G}$: $V \rightharpoonup \text{Geno}(\mathbf{A})$, that associates a genotype to (some of the) members of the pedigree. The domain, $dom(\mathcal{G})$, of the function is referred to as the *set of genotyped members* of the pedigree. If $\mathbf{m} \in dom(\mathcal{G})$ we say that $\mathbf{m}$ is genotyped (at that marker).

Remark 2. Note that in the definition above, a genotype assignment assigns an *unordered* pair of elements to members of the family. This indicates that the phase of the alleles is unknown. If a pedigree member is *homozygous* at a given locus, i.e. the same allele is assigned to both of its bits, the function $\mathcal{G}$ returns a singleton set.

Compatibility: A founder allele assignment $\mathcal{F}$ is said to be *compatible* with the genotype assignment $\mathcal{G}$ and the inheritance pattern i if the following holds:

$$\forall \mathbf{n} \in dom(\mathcal{G}) \ . \ \mathcal{F}(\mathsf{Source}^{\mathsf{i}}(\mathbf{n})) = \mathcal{G}(\mathbf{n}) \ .$$

We let $Comp_P(\mathsf{i}, \mathcal{G})$ denote the set of such founder allele assignments. If $Comp_P(\mathsf{i}, \mathcal{G}) \neq \emptyset$ then i is said to be compatible with $\mathcal{G}$.

Algorithm for the Single Point Probability Calculations:

Remark 3. In what follows if $f, g : A \longrightarrow \mathbb{R}$ we write $f \sim g$ (or sometimes $f(a) \sim_a g(a)$) iff there is a constant K such that for all $a \in A$, $f(a) = K \cdot g(a)$.

A genotype (at a fixed marker) can be represented as a set of pairs of the form $(\mathbf{n}, \{a_1, a_2\})$, where $\mathbf{n}$ is a member of the family and a_1 and a_2 are alleles. Intuitively such a pair is supposed to indicate that these two alleles have to be assigned to the allelic positions of $\mathbf{n}$, $(\mathbf{n}, m)$ and $(\mathbf{n}, p)$; we just do not know which is assigned to which as phase cannot be decided directly. If the inheritance pattern i is known, we may find out from which founder positions the alleles of the positions of $\mathbf{n}$ are originated. Therefore, in this case we may consider the genotype assignment as an implicit specification of a founder allele assignment as the allelic positions of $\mathbf{n}$ must contain the same allele values as the founder positions they came from.

In what follows, we will describe the algorithm of the software tool Allegro at De-Code Genetics, Reykjavik. More details about the algorithm may be found in [10].

Before we present the algorithm, we will explain the notation it relies on. In the head of the algorithm we have the set I_P, the set of all inheritance patterns over P. The algorithm takes as input a pedigree P and genotype assignment $\mathcal{G}$ and return a mapping that, to each inheritance pattern $\mathsf{i} \in \mathsf{I}_P$, assigns a pair $(\mathcal{S}, \mathcal{P})$, that can be considered as an implicit representation of the set of founder allele assignments which are compatible with P and $\mathcal{G}$ given i; if the input is incompatible the algorithms return the element $\bot$. We refer to elements of the type described above as implicit founder allele assignment (ifa) and the set of such elements is denoted by IFA_P.

As each pair of the form $(\mathcal{S}, \mathcal{P})$ specifies a unique set of founder allele assignments, their probabilities can be calculated from the frequency distribution of the alleles. This will be explained later in this section.

The components $\mathcal{S}$ and $\mathcal{P}$ represent the fully specified part of the assignment and the partially specified one respectively. The set $\mathcal{S}$ consists of elements of the form (f, a), where f is an allelic position of a founder and a is an allele. This pair represents the set of all founder allele assignments that map f into a. The second set $\mathcal{P}$ consists of elements of the form $(\{f_1, f_2\}, \{a_1, a_2\})$ (for short we often write $(f_1 f_2, a_1 a_2)$ instead), which represents the set of assignments that either map f_1 to a_1 and f_2 to a_2, or f_1 to a_2 and f_2 to a_1. We let $dom(\mathcal{S})$ and $dom(\mathcal{P})$ denote the set of positions that occur in the sets $\mathcal{S}$ and $\mathcal{P}$ respectively. The interpretation described above can be expressed formally by the function Interpret as follows.

Definition 1. *The function* Interpret *is defined as follows:*

$$\text{Interpret}(f, a) = \{\mathcal{A} | \mathcal{A}(f) = a\},$$
$$\text{Interpret}(f_1 f_2, a_1 a_2) = \{\mathcal{A} | \mathcal{A}(f_1) = a_1 \text{ and } \mathcal{A}(f_2) = a_2,$$
$$\text{or } \mathcal{A}(f_1) = a_2 \text{ and } \mathcal{A}(f_2) = a_1\},$$
$$\text{Interpret}(\mathcal{X}) = \bigcap\nolimits_{\alpha \in \mathcal{X}} \text{Interpret}(\alpha) \text{ for } \mathcal{X} = \mathcal{S}, \mathcal{P},$$
$$\text{Interpret}(\mathcal{S}, \mathcal{P}) = \text{Interpret}(\mathcal{S}) \cap \text{Interpret}(\mathcal{P}),$$
$$\text{Interpret}(\perp) = \emptyset.$$

$\gamma_1, \gamma_2 \in \text{IFA}_P$ *are said to be* equivalent, $\gamma_1 \sim \gamma_1$, *iff* $\text{Interpret}(\gamma_1) = \text{Interpret}(\gamma_2)$.

Lemma 1. *If* $(\mathcal{S}, \mathcal{P}) \sim (\mathcal{S}', \mathcal{P}')$ *then* $dom(\mathcal{S}) \cup dom(\mathcal{P}) = dom(\mathcal{S}') \cup dom(\mathcal{P}')$.

Definition 2. *The* ifa $(\mathcal{S}, \mathcal{P})$ *is said to be* fully reduced *if it is either* $\perp$ *or following is satisfied:*

- $dom(\mathcal{S}) \cap dom(\mathcal{P}) = \emptyset$ *and*
- *if* $(\mathcal{S}, \mathcal{P}) \sim (\mathcal{S}', \mathcal{P}')$ *then* $dom(\mathcal{S}') \subseteq dom(\mathcal{S})$.

Lemma 2.   – *For each element* $\gamma \in \text{IFA}_P$ *there exists exactly one fully reduced* $\eta \in \text{IFA}_P$, *which is equivalent to it. We refer to* η *as the* normal form *for* γ.
  – $\gamma \in \text{IFA}_P$ *is incompatible iff its normal form is* $\perp$.

Algorithm 1 BUILDFORFOUNDERS(**in:** $\mathcal{P}, \mathcal{G}$; **out:**$\mathcal{P}'$)

$\mathcal{P}' \leftarrow \mathcal{P}$
for $n \in dom(\mathcal{G}) \cap F$ **do**
    $\mathcal{P}' \leftarrow \mathcal{P}' \cup \{\{(\mathbf{n}, m), (\mathbf{n}, p)\}, \mathcal{G}(\mathbf{n}))\}$
end for

The algorithm by Gudbjartsson et al safes both time and space by discovering incompatibilities in subpedigrees. The theoretical considerations behind this approach can be described in the following lemma, where a subpedigree of a pedigree has the expected meaning.

Lemma 3. *If* $(P, \mathcal{G}, i)$ *is compatible,* $P' = (V', F', \phi')$ *is a subpedigree of* P *and* i' *and* $\mathcal{G}'$ *are* i *and* $\mathcal{G}$ *restricted to* $N' \cup F'$ *respectively, then* $(P', \mathcal{G}', i')$ *is also compatible.*

Intuitively Lemma 3 says that any substructure of a compatible structure of the form $(P, \mathcal{G}, i)$ must be compatible too. Turning this reasoning the other way around we get that if such a structure is incompatible, then all extensions of it remain incompatible.

The algorithm SINGLEPOINT works by first calling the algorithm BUILDFOR FOUNDERS that investigates the initial substructure only consisting of the founder positions and the corresponding genotype assignment. Then it calls the recursive algorithm TRAVERSETREE that gradually extends this structure by, at each recursive call, adding a new member to the pedigree and to extend the genotype assignment according to the original input. At the same time it extends the inheritance pattern in both possible, i.e by investigating both the extension by adding 0 and 1 to all existing patterns. For this to work we preassume an order o of the non founders of the input pedigree that respects the family hierarchy. More precisely we assume that if $o(\mathbf{n1}) < o(\mathbf{n2})$ then $(\mathbf{n1}, \mathbf{n2}) \in (\phi_p \cup \phi_m)^+$. Furthermore an inheritance vector is represented by a string $v \in \{0, 1\}^{2K}$, where K is the number of non founders of the family. If $v = b_1^p b_1^m \cdots b_K^p b_K^m$ this can be interpreted by $v(\mathbf{n}j, p) = b_j^p$ and $v(\mathbf{n}j, m) = b_j^m$. The algorithm of Gudbjartsson et al is given in Algorithms 1–4.

Algorithm 2 SINGLEPOINT(**in**: $\mathcal{G}, P$;**out**: $\Delta \in \mathsf{I}_P \longrightarrow \mathsf{IFA}_P$)

$(\mathcal{S}, \mathcal{P}) \leftarrow$ BUILDFORFOUNDERS$(\mathcal{G}, P)$
TRAVERSETREE$(\varepsilon, (\mathcal{S}, \mathcal{P}), \mathcal{G}, P)$

Algorithm 3 TRAVERSETREE($\mathbf{in} : v, (\mathcal{S}, \mathcal{P}), \mathcal{G}, P; \mathbf{out} : \Delta$)

---

if $|v| = |N|$ **then**
  return$(v, (\mathcal{S}, \mathcal{P}))$
else
  for $(b_p, b_m) \in \{0, 1\}^2$ **do**
    $v' \leftarrow v b_p b_m$
    $\gamma \leftarrow (\mathcal{S}, \mathcal{P})$
    $j \leftarrow |v'|$
    if $n_j \in dom(\mathcal{G})$ **then**
      $\gamma \leftarrow$INSERT(Source$^{v'}(\mathbf{n}), \mathcal{G}(\mathbf{n}), \gamma)$
      if $\gamma = \bot$ **then**
        return$(v', \bot)$
      else
        TRAVERSETREE$(v', \gamma, \mathcal{G}, P)$
      end if
    else
      TRAVERSETREE$(v', \gamma, \mathcal{G}, P)$
    end if
  end for
end if

The algorithm TRAVERSETREE calls the algorithm INSERT, that takes a pair of founder allelic positions, a genotype and a reduced ifa and returns a reduced ifa. To prove the correctness of the algorithm SINGLEPOINT we show that it satisfies the invariance conditions states in the following lemma.

Lemma 4. *Let P_l and $\mathcal{G}_l$ denote the subpedigree and genotype assignment obtained by only considering the set, $\{\mathbf{n}1, \dots, \mathbf{n}l\}$, of pedigree members. Let furthermore* Interpret$_{P_l}$ *be the interpretation function w.r.t. P_l. Then we have:*

– *If γ is fully reduced, then* Insert(ff', aa', γ) *is also fully reduced.*
– *If* Interpret$_{P_l}(\gamma) = Comp_{P_l}(v, \mathcal{G}_l)$*, then*

Algorithm 4 INSERT($\mathbf{in}: \{f, f'\}, \{a, a'\}, (\mathcal{S}, \mathcal{P}); \mathbf{out}: \gamma$)

$\gamma \leftarrow (\mathcal{S}, \mathcal{P})$
if $f = f'$ and $a = a'$ **then**
  if $(f, a) \in \mathcal{S}$ **then**
    do nothing
  else if $(f\phi, a\alpha) \in \mathcal{P}$ for some ϕ, α **then**
    remove $(f\phi, a\alpha)$ from $\mathcal{P}$
    add (f, a) and (ϕ, α) to $\mathcal{S}$
  else if $f \notin \mathcal{S} \cup \mathcal{P}$ **then**
    add (f, a) to $\mathcal{S}$
  else
    $\gamma \leftarrow \bot$
  end if
else if $f \neq f'$ **then**
  if $(f, a), (f', a') \in \mathcal{S}$ **then**
    do nothing
  else if $(f, a) \in \mathcal{S}, (f'\phi, a'\alpha) \in \mathcal{P}$ for some ϕ, α **then**
    remove $(f'\phi, a'\alpha)$ from $\mathcal{P}$
    add $(f', a'), (\phi, \alpha)$ to $\mathcal{S}$
  else if $(f\phi, a\alpha), (f'\phi', a'\alpha') \in \mathcal{P}$ **then**
    remove $(f\phi, a\alpha), (f'\phi', a'\alpha')$ from $\mathcal{P}$
    add $(f, a), (f', a'), (\phi, \alpha), (\phi', \alpha')$ to $\mathcal{S}$
  else if $(ff', aa') \in \mathcal{P}'$ **then**
    do nothing
  else if $(f, a) \in \mathcal{S}, f' \notin \mathcal{S} \cup \mathcal{P}$ **then**
    add (f', a') to $\mathcal{S}$
  else if $(f\phi, a\alpha) \in \mathcal{P}, f' \notin \mathcal{S} \cup \mathcal{P}$ **then**
    remove $(f\phi, a\alpha)$ from $\mathcal{P}$
    add $(f, a), (\phi, \alpha), (f', a')$ to $\mathcal{S}$
  else if $f, f' \notin \mathcal{S} \cup \mathcal{P}$ and $a \neq a'$ **then**
    add (ff', aa') to $\mathcal{P}$
  else if $f, f' \notin \mathcal{S} \cup \mathcal{P}$ and $a = a'$ **then**
    add $(f, a), (f', a')$ to $\mathcal{S}$
  else
    $\gamma \leftarrow \bot$
  end if
else
  $\gamma \leftarrow \bot$
end if

- Interpret$_{P_{l+1}}(\gamma) = Comp_{P_{l+1}}(vb_p b_m, \mathcal{G}_{l+1})$ *for* $b_p, b_m \in \{0, 1\}$, *if* $\mathbf{n} \notin dom(\mathcal{G})$.
- Interpret$_{P_{l+1}}(\text{Insert}(\text{Source}^{vb_p b_m}(\mathbf{n}), \mathcal{G}(\mathbf{n}), \gamma)) = Comp_{P_{l+1}}(vb_p b_m, \mathcal{G}_{l+1})$ *for* $b_p, b_m \in \{0, 1\}$, *if* $\mathbf{n} \in dom(\mathcal{G})$.

The following corollary states exactly the correctness of the algorithm and follows directly from Lemma 4.

Corollary 1. – *If he algorithm* SINGLEPOINT *outputs* (v, γ), *where* $\gamma \neq \perp$, *then* $|v| = |N|$ *and* Interpret$(\gamma) = Comp_P(v, \mathcal{G})$.

– *If it outputs* $(v, \perp)$, *then* $(P_{|v|}, \mathcal{G}_{|v|}, v)$ *and all its extensions are incompatible, (i.e. the inheritance vector obtained as* st *is incompatible with* P *and* $\mathcal{G}$ *for all* $t \in \{0, 1\}^{2(|V|-|v|)})$.

Probability Calculations for the Single Point Case. To calculate the single point probabilities $p(v|\mathcal{G})$ for a given inheritance vector v, we first note that by Bayes' theorem

$$p(v|\mathcal{G}) = \frac{p(\mathcal{G}|v)}{p(v)} p(\mathcal{G}) = K p(\mathcal{G}|v),$$

where $K = \frac{p(\mathcal{G})}{p(v)}$ is independent of v. Therefore $p(v|\mathcal{G}) \sim_v p(\mathcal{G}|v)$ and it is sufficient to calculate $p(\mathcal{G}|v)$ for all v and replace γ in the output (v, γ) of the algorithm by the result. As explained earlier, for a given v, the pedigree P will have the genotype given by $\mathcal{G}$ exactly for founder allele assignments from the set $Comp_P(v, \mathcal{G})$. This set, in turn, is uniquely decided by the pair $(\mathcal{S}, \mathcal{P})$ that is associated to v by the algorithm above when it is run on $\mathcal{G}$ as input. In the calculations of $p(\mathcal{S}, \mathcal{P})$ we can assume that the allele frequency is independent of the founders. Thus, if we let π denote the frequency distribution over A in the given population, this probability can be obtained as follows:

– $p(g, a) = \pi(a)$,
– $p(f_1 f_2, a_1 a_2) = 2\pi(a_1)\pi(a_2)$,
– $p(\mathcal{X}) = \Pi_{\alpha \in \mathcal{X}} p(\alpha)$ for $\mathcal{X} = \mathcal{S}, \mathcal{P}$,
– $p(\mathcal{S}, \mathcal{P}) = p(\mathcal{S})p(\mathcal{P})$,
– $p(\perp) = 0$.

4   The Multi Point Case

Marker Maps and Multi Point Genotypes: In the single point analysis we calculated the probability distribution over all inheritance vectors at each marker given the genotype information at that marker but independent of the genotype information at all other markers.

The aim of this section is to show how the multi point probabilities can be derived from the single point probabilities and the recombination fraction. For this purpose we need a *marker map* (of size M) which is a sequence of the form

$$\mathcal{M} = ((\mathbf{A}_1, \pi_1), \theta_1, \ldots, (\mathbf{A}_{M-1}, \pi_{M-1}), \theta_{M-1}, (\mathbf{A}_M, \pi_M)),$$

where $\mathbf{A}_i$ is the set of alleles for marker i, π_i is a frequency distribution over $\mathbf{A}_i$ for $i \leq M$ and θ_i is the recombination fraction between markers i and $i + 1$ for $i \leq M - 1$.

Furthermore a *multi point genotype information*, $\mathcal{G}$, for the marker map $\mathcal{M}$ is an M-dimensional vector $\mathcal{G} = (\mathcal{G}_1, \mathcal{G}_2, \ldots, \mathcal{G}_M)$, where $\mathcal{G}_i$ is a single point genotype information over $\mathbf{A}_i$ for $i \leq M$. For $i > 1$ we let $\mathcal{G}_i^- = (\mathcal{G}_1, \ldots, \mathcal{G}_{i-1})$, and for $i < M$ we let $\mathcal{G}_i^+ = (\mathcal{G}_{i+1}, \ldots, \mathcal{G}_M)$.

We recall that the recombination fraction θ_i tells us that if an individual n gets, say, his maternal allele from his grandfather at marker i, then he gets the maternal allele from his grandmother at marker $i + 1$ with probability θ_i. This means that the value of i on the position $(\mathbf{n}, m)$ changes from 0 at marker i to 1 at marker $i + 1$ with probability θ_i. We also note that the probability that a recombination takes place for one members of the pedigree is independent of recombinations taking place in the other members. Thus the probability that the inheritance vector changes from v to w between the markers i and $i + 1$ is only dependent on the Hamming distance between v and w, indicated by $|v - w|$, and is given by $\theta_i^{|v-w|} \bar{\theta}_i^{|N-v-w|}$, where $\bar{\theta} = 1 - \theta$. This result will be used in the following calculations.

The Lander-Green Algorithm: Next we introduce the random variables $X_1, \ldots, X_M$ where X_k takes the value v if the inheritance vector at marker k is given by v (assuming a fixed bit order). The single point probability calculations amount to calculating $p(X_k = v | \mathcal{G}_k)$ for all v for $k = 1, \ldots, M$. The multi point calculations at the other hand aim at calculating $p(X_k = v | \mathcal{G})$ for all v for $k = 1, \ldots, M$ from the single point calculations and the recombination fractions. To obtain this we assume the following properties.

MARKOV PROPERTY (MP): For $1 < k < M$,

$$p(X_{k+1} = v | X_k = w, X_{k-1} = u) = p(X_{k+1} = v | X_k = w),$$
$$p(X_{k+1} = v | X_k = w, \mathcal{G}_k^-) = p(X_{k+1} = v | X_k = w).$$

LINKAGE EQUILIBRIUM (LE): For $1 < k < M$,

$$p(\mathcal{G} | X_{k-1} = v_1, X_k = v_2, X_{k+1} = v_3) =$$
$$p(\mathcal{G}_k^- | X_{k-1} = v_1) p(\mathcal{G}_k | X_k = v_2) p(\mathcal{G}_k^+ | X_{k+1} = v_3).$$

The Markov Property says that the probability of observing the inheritance vector v at marker $k + 1$, given the inheritance vector at marker k, is independent of both the value of the inheritance vector and the genotype at any marker to the left of marker k.

The Linkage Equilibrium in more general terms than it appears here, says that the frequency of random *haplotypes* (i.e. a sequence of alleles that occur at consecutive markers) is the product of the frequency of the underlying haplotypes [16].

For both MP and LE we assume right hand counter parts (i.e. with loci with lower index replaced by loci with higher index), which are not given here.

As before, in what follows if $f, g : A \longrightarrow \mathbb{R}$ we write $f \sim g$ (or sometimes $f(a) \sim_a g(a)$) iff there is a constant K such that for all $a \in A$, $f(a) = K \cdot g(a)$.

The following theorem is the base for the multi point calculation presented in this paper; the proof follows from MP and LE and a straight forward application of Bayes' Theorem.

Theorem 1.

1. $p(X_k = v | \mathcal{G}) \sim_v p(X_k = v | \mathcal{G}_k^-) p(X_k = v | \mathcal{G}_k) p(X_k = v | \mathcal{G}_k^+).$
2. *(a)* $p(X_k = v | \mathcal{G}_k^-) \sim_v$
$$\sum_w p(X_{k-1} = w | \mathcal{G}_{k-1}^-) p(X_{k-1} = w | \mathcal{G}_{k-1}) p(X_k = v | X_{k-1} = w),$$
   (b) $p(X_k = v | \mathcal{G}_k^+) \sim_v$
$$\sum_w p(X_{k+1} = w | \mathcal{G}_{k+1}^+) p(X_{k+1} = w | \mathcal{G}_{k+1}) p(X_{k+1} = w | X_k = v).$$

Let us now simplify the notation above and define

- $mp_k(v) = p(X_k = v|\mathcal{G})$ (the multi point probability of v at marker k),
- $sp_k(v) = p(X_k = v|\mathcal{G}_k)$ (the single point probability of v at marker k),
- $l_k(v) = p(X_k = v|\mathcal{G}_k^-)$ (the left contribution for v at marker k),
- $r_k(v) = p(X_k = v|\mathcal{G}_k^+)$ (the right contribution for v at marker k) and
- $T_k(w, v) = p(X_{k+1} = v|X_k = w)$ (the transition probability between w and v at marker k).

As we mentioned earlier, the probability of observing a recombination from marker k to marker $k + 1$ for a given individual is θ_k. As the recombination events are independent between individuals, we get that $T_k(w, v) = \theta_k^{|w-v|}\overline{\theta}_k^{|N|-|w-v|}$, where $|N|$ is the size of the set N and $|v - w|$ is the Hamming distance (the number of different bits) between w and v. Often we will write $t_k(v - w)$ instead of $T(w, v)$, emphasizing the fact that the output of T only depends on the Hamming distance of the inputs. In this notation Theorem 1 becomes:

1. $mp_k(v) \sim_v l_k(v) \cdot sp_k(v) \cdot r_k(v)$ (or equivalently $mp_k \sim l_k \cdot sp_k \cdot r_k$) for $0 \leq k \leq M$,
2. (a) $l_k(v) \sim_v \sum_w l_{k-1}(w) \cdot sp_{k-1}(w) \cdot T_{k-1}(w, v)$ (or equivalently $l_k \sim_v (l_{k-1} \cdot sp_{k-1}) * t_{k-1}$)
   for $1 < k \leq M$, and $l_1(v) = 1$,
   (b) $r_k(v) \sim_v \sum_w r_{k+1}(w) \cdot sp_{k+1}(w) \cdot T_k(w, v)$ (or equivalently $r_k \sim_v (r_{k+1} \cdot sp_{k+1}) * t_k$)
   for $0 \leq k < M$, and $r_1(v) = 1$.
   where $f * g$ (the convolution of f with g) is defined by $(f * g)(v) = \sum_w f(w) \cdot g(v - w)$.

From this we derive an iterative algorithm, the Lander-Green algorithm [17], that calculates mp_k for each marker k from sp_j and θ_j for all j. The details of the algorithm are given in Algorithm 5.

Algorithm 5 LanderGreen (**in** : $(sp_1, \ldots, sp_M), (\theta_1, \ldots, \theta_{M-1})$, **out** : $(mp_1, \ldots, mp_M)$

$l_1 \leftarrow 1$
for $i = 1$ to $M - 1$ **do**
  $l_{i+1} \leftarrow (l_i \cdot sp_i) * t_i$
end for
$r_M \leftarrow 1$
for $i = M$ **to** 2 **do**
  $r_{i-1} \leftarrow (r_i \cdot sp_i) * t_{i-1}$
end for
for $i = 1$ to M **do**
  $mp_i \leftarrow l_i \cdot sp_i \cdot r_i$
end for
return $(mp_1, \ldots, mp_M)$

The running time of the Lander-Green algorithm is of the order $O(M \cdot g(n))$, where M is the number of markers (loci) under investigation and $g(n)$ is the time complexity of the algorithm used to calculate the convolution operator $*$ for a pedigree of size n, where $n = 2|N|$ and N is the set of non founders of the pedigree. It is therefore clear that the allover performance of the algorithm depends entirely on $f(n)$. What remains of the paper is devoted to describing the calculations of the convolution $*$.

The Indury-Elston Algorithm: When the meaning is clear from the context, we denote $f * t$ by f^* and simply talk about the contribution of f. Furthermore, in the following theorem we let $\bar{\theta} = 1 - \theta$ and let for $w_1 w_2 \in \mathbb{B}^r$ we define $f_{w_1}(w_2) = f(w_1 w_2)$ and $f_{w_1}^*(w_2) = (f_{w_1})^*(w_2)$.
　　The following theorem is a straight forward consequence of Theorem 1.

Theorem 2. *We have that*

$$f^*(v) = f_\varepsilon^*(v) \, for \, v \in \mathbb{B}^r$$

where

$$f_w^*(0v) = \bar{\theta} f_{w0}^*(v) + \theta f_{w1}^*(v) \, for \, w0v \in \mathbb{B}^r,$$

$$f_w^*(1v) = \theta f_{w0}^*(v) + \bar{\theta} f_{w1}^*(v) \, for \, w1v \in \mathbb{B}^r \, and$$

$$f_w^*(\varepsilon) = f(w) \, for \, w \in \mathbb{B}^r$$

and ε is the empty string.

From Theorem 2 we derive the algorithm IDURYELSTON, a slight reformulation of the Idury-Elston algorithm of [13], that is given in Algorithm 6. It runs in $O(n2^n)$ and space $O(2^n)$ where $n = 2|N|$.

Algorithm 6 IduryElston(**in** : $f(w), w \in \mathbb{B}^n$ **out** : $f^*(v), v \in \mathbb{B}^n$)

---

for $w = 0^n$ to 1^n **do**
　$a_w \leftarrow f(w)$
end for
for $j = 0$ to $n - 1$ **do**
　for $w = 0^j$ to 1^j **do**
　　for $v = 0^{n-j-1}$ to $= 0^{n-j-1}$ **do**
　　　$x \leftarrow a_{w0v}$
　　　$y \leftarrow a_{w1v}$
　　　$a_{w0v} \leftarrow \bar{\theta} x + \theta y$
　　　$a_{w1v} \leftarrow \theta x + \bar{\theta} y$
　　end for
　end for
end for
return$(a_{0^n}, \ldots, a_{1^n})$

We complete this discussion by stating the following theorem that allows for reduction of the inheritance space.

Theorem 3. *If $\phi : \mathbb{B}^n \longrightarrow \mathbb{B}^n$ is a bijection and p a probability distribution over $\mathbb{B}^n$, such that $p \circ \phi = p$ and $|\phi(v) - \phi(w)| = |v - w|$ for all $v, w \in \mathbb{B}^n$ then $(p * t) \circ \phi = p * t$.*

In words Theorem 3 says that if ϕ is a bijection of the inheritance space that preserves a probability distribution over $\mathbb{B}^n$ (e.g. the single point probability of the vectors) and the distances between vectors, then it preserves the probability given by $p * t$ too. Consequently it also preserves the multi point probabilities.

5 Implementation Based on Multi Terminal Binary Decision Diagrams

Binary Decision Diagrams (BDDs) [5,3] were originally introduced as a compact representation of Boolean functions to reason about digital circuits. Later their use has been extended to other disciplines but mainly software verification. BDDs provide a compact symbolic representation of computational problems that often offer a practical solution to even NP-hard problems. The Multi Terminal BDDs (sometimes called Algebraic BDDs) are extensions of the original BDDs as they allow for encoding of functions from Booleans into any finite set of values. In what follows we will give a short survey of MTBDDs. For a more detailed account on the subject see for instance [4,7]. We assume that *Val*, ranged over by v, v' etc., is a finite set of values and *Var*, range over by x, y etc., is a finite set of variables.

1. A *Multi Terminal BDD (MTBDD)* is a single rooted directed acyclic graph with
   - a set of terminal nodes T (ranged over by t) of out-degree zero labelled over *Val*; the value label is given by $val(t)$.
   - a set of variable nodes D (ranged over by d, d_1, d' etc.) of out-degree two, labelled over *Var*. The variable label is given by $var(d)$. The two outgoing edges are labelled with 0 and 1. The node the 0-arc leads to is given by $L(d)$ (left-hand side of d) and the node the 1-arc leads to is given by $R(d)$ (the right hand side of d). We let d range over the set $Nodes = D \cup T$ too.
2. An MTBDD is ordered (with respect to an order e over *Var*) (O-MTBDD) if for all $d \in Nodes$, either $d \in T$ or both
   (a) $L(d) \in T$ or $e(var(L(d))) < e(var(d))$ and
   (b) $R(d) \in T$ or $e(var(R(d))) < e(var(d))$
   hold.
3. An O-MTBDD is *Reduced* (RO-MTBDD) if following holds:
   - (**Uniqueness**) $var(d_1) = var(d_2)$, $L(d_1) = L(d_2)$ and $R(d_1) = R(d_2)$ implies $d_1 = d_2$, and $val(t_1) = val(t_2)$ implies $t_1 = t_2$.
   - (**Non-redundant tests**) $L(d) \neq R(d)$.
4. The function **make** (x, d_l, d_r) creates a new MTBDD with d_l as a left hand node, d_r as a right hand node and where $var(\textbf{make }(x, d_l, d_r)) = x$. The function **delete** (d) removes d and makes the space it took up available. We also assume the binary operations $+$ and the unary multiplication by a constant $c \cdot \_$ on MTBDDs.

From now on we assume that all MTBDDs are reduced and ordered and refer to them simply as MTBDDs.

The MTBDDs have the key property that for a fixed variable order, each function that maps vectors of Booleans to Val has exactly one MTBDD representation. The size of the MTBDD depends heavily on the order of the variables (here the bits) and therefore we want to be able to investigate different orders in our implementation.

Now we will describe how we implement our algorithm in terms of the MTBDDs. The first step towards this is to assign a variable to each position or bit in the family. Then we change the single point algorithm in such a way that it returns a MTBDD that represents the probability distribution over inheritance vectors. (This is straight forward and will not be described here.) In particular the vectors with probability 0 are omitted which in many cases reduces the resulting MTBDD considerably. (As pointed out earlier, the size of the MTBDD depends heavily on the bit order and therefore in our implementation we allow for different orders).

The algorithm M in Algorithm 7 gives a sketch of a version of the algorithm that does not cater for founder and founder couple reduction and it works as follows: It takes as an input a MTBDD d_{in} that encodes the single point probability (sp) and outputs a new MTBDD d_{out} that encodes the convolution of the the probability distribution with the transition probability function ($sp * t$). If d_{in} is a terminal, M outputs d_{in} directly. Otherwise M is called recursively on $L(d_{in})$ and $L(d_{out})$, the multiplications and additions are performed on the result as described in Theorem 2 and the resulting MTBDD is created by calling **make** . As soon as an MTBDD is not needed any more in the calculations, it is deleted by calling **delete** and the space it took up can be reused.

Both the worst case time and space complexity of the algorithm M is $O(n2^{2n})$.

6   Reduction of the Inheritance Space

In this section we will show how the inheritance space can be reduced by recognizing some symmetries, i. e. by taking advantages of the fact that certain sets of inheritance patterns must have exactly the same probability independent of both markers and geno-types given some mild restrictions on the latter. In particular we will focus on the so

---

Algorithm 7 M(**in** : d_{in}; **out** : d_{out})

---

if $d_{in} \in T$ **then**
    $d_{out} \leftarrow d_{in}$
else
    $x \leftarrow var(d_{in})$
    $d_1 \leftarrow$ M($L(d_{in})$)
    $d_2 \leftarrow$ M($R(d_{in})$)
    delete d_{in}
    $d'_1 \leftarrow \bar{\theta} \cdot d_1,\ d'_2 \leftarrow \theta \cdot d_2$
    $d_l \leftarrow d'_1 + d'_2,$ **delete** d'_1, d'_2
    $d'_1 \leftarrow \theta \cdot d_1,$ **delete** d_1
    $d'_2 \leftarrow \bar{\theta} \cdot d_2,$ **delete** d_2
    $d_r \leftarrow d'_1 + d'_2,$ **delete** d'_1, d'_2
    $d_{out} \leftarrow$ **make** (x, d_l, d_r)
end if

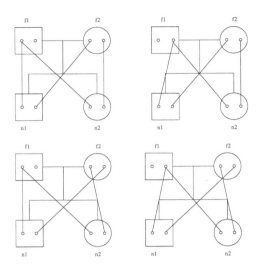

Fig. 3. Equivalent inheritance patterns with respect to founder reduction

called *founder reductions* [14] and *founder couple reductions* [10]. The main challenge of this paper is to provide a description of these reductions in terms of MTBDDs so they can be directly incorporated into the MTBDD based algorithms. In what follows we will describe the main ideas of this translation. For a more detailed description we refer to the full version of the paper [12].

Founder Reduction: In the founder reduction we make use of the fact that there is no information available for the parents of the founders and therefore interchanging references to them, or equivalently to the founders parental positions, should not change the probability of the resulting inheritance pattern.

Figure 3 shows all equivalent inheritance patterns for the founder reduction with respect to the founders $\mathbf{f}1$ and $\mathbf{f}2$.

To give a more precise description of the founder reduction, we need to introduce a few concepts. The position $(\mathbf{m}', b)$ is said to be a *direct descent of* $\mathbf{m}$ if $\phi_b(\mathbf{m}') = \mathbf{m}$ (i.e. if the allele sitting in the position $(\mathbf{m}', b)$ is inherited directly from $\mathbf{m}$). The set of direct descents of $\mathbf{m}$ is referred to as $Des(\mathbf{m})$. Thus two inheritance patterns are equivalent with respect to a founder $\mathbf{f}$ if one of them is derived by flipping the values of exactly the bits corresponding to $Des(\mathbf{f})$. We note that the sets $Des(\mathbf{f})$ for $\mathbf{f} \in F$ are mutually disjoint. Two inheritance patterns are founder reduction equivalent if, for some set $F' \subseteq F$ of founders, one of them is derived from the other by flipping all the bits of $Des(F')$. This implies that each equivalence class contains exactly $2^{|F|}$ inheritance patterns where $|F|$ is the number of founders in the pedigree. Therefore we have $2^{2|N|-|F|}$ different equivalence classes for founder reduction. Sometimes we refer to the sets $Des(\mathbf{f})$ as the *founder reduction block* associated to $\mathbf{f}$.

To reduce the inheritance space, we can pick exactly one representative for each equivalence class and use only this in the calculations of the convolutions for the whole class. (That this is possible, follows from Theorem 3.) We do this systematically by, for each $\mathbf{f} \in F$, choosing exactly one $(\mathbf{m}, b) \in Des(\mathbf{f})$. It is not difficult to show that for each equivalence class there is exactly one representative, where the inheritance pattern or vector has the value 0 on this set. We refer to this set of positions as Fix (the set of fixed bits for founder reduction). We call these representatives the *canonical representatives with respect to* Fix. If we express the canonical representatives as an MTBDD, all the bits defined on Fix are fixed to 0 and can be omitted. This will give a much smaller MTBDD with maximum $2|N| - |F|$ bits. This is what is done in the real implementation of the algorithm. However, to simplify the description of the algorithm, we assume that the bits in Fix are still there but that the corresponding node only has a left node but not a right node. This has little influence on the running time and the space requirements.

A sketch of the multi point algorithm with founder reduction included (MF) is given in Algorithm 8. It assumes that the bit order satisfies the following conditions:

- It respects the blocks (the blocks do not overlap)
- The fixed bits come first in each block.

In the algorithm we assume the following background knowledge (i.e. is not given as a parameter in the input) :

- A function pos that reads from each node the position it represents.
- The set Fix of fixed positions.
- A function $flip$ that associates, to each position in Fix, the set of positions (descendants of founders) that have to be flipped by the algorithm in the corresponding block.

Algorithm 8 MF(**in** : d_{in}, Fl_{in}; **out** : d_{out})

---

  if $d_{in} \in T$ **then**
    $d_{out} \leftarrow d_{in}$
  else
    if $pos(d_{in}) \in Fix$ **then**
      $d_1 \leftarrow \text{MF}(L(d_{in}), \emptyset)$
      $d_2 \leftarrow \text{MF}(L(d_{in}), flip(pos(d_{in})))$
    else
      if $pos(d_{in}) \in Fl_{in}$ **then**
        $d_1 \leftarrow \text{MF}(R(d_{in}), Fl_{in})$
        $d_2 \leftarrow \text{MF}(L(d_{in}), Fl_{in})$
      else
        $\vdots$
        (as before)
        $\vdots$
        $d_{out} \leftarrow Make(x, d_l, d_r)$
      end if
    end if
  end if

The founder reduced version of the algorithm, MF, given in Algorithm 7 works as follows: It takes as an input a founder reduced MTBDD d_{in} (provided by the single

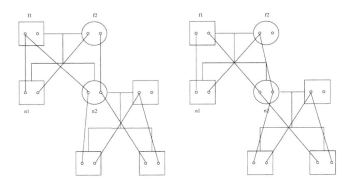

Fig. 4. Two equivalent inheritance patterns according to founder-couple reduction

point phase) and a set of positions to be flipped in the current block. As before it outputs d_{out}. Again as before, if d_{in} is a terminal, MF outputs d_{in} directly. Otherwise, if $pos(d_{in}) \in Fix$, then we are starting on a new block (recall that the fixed bits are placed at the beginning of a block). As before, algorithm is called recursively on $L(d_{in})$. On the other hand, as the bit of the current position is fixed to 0, $R(d_{in})$ does not exist. Instead we call the algorithm again recursively on $L(d_{in})$ (i.e. we flip the current bit) and give as a parameter the set of bits that should be flipped further within the new block ($flip(pos(d_{in}))$). Otherwise, if $pos(d_{in}) \in Fl_{in}$ we flip the bits both on the left (by calling MF on $R(d_{in})$) and the right substructure (by calling MF on $L(d_{in})$). Otherwise we continue as before.

Founder Couple Reduction: The founder couple reduction is based on the observation that ungenotyped parents that form a founder couple are indistinguishable.

Figure 4 shows two inheritance patterns that are equivalent according to the founder couple reduction induced by the founder couple $\mathbf{f}1, \mathbf{f}2$. As can be seen from Figure 4, two inheritance patterns are founder couple equivalent if one of them is derived from the other by

- swapping the maternal bits of children of the couple,
- swapping the paternal bits of children of the couple and
- flipping the bits of the direct descendants of the children of the couple (referred to as the *second descendants* of the couple, $SecDes(\mathbf{f}1, \mathbf{f}2)$.

On top of this the usual founder reduction can be applied and we get equivalent inheritance patterns by flipping the bits of either $Des(\mathbf{f}1)$ or $Des(\mathbf{f}2)$ or both. For each founder couple we can now fix one bit from $SecDes(\mathbf{f}1, \mathbf{f}1)$ and, as before, one from each of $Des(\mathbf{f}1)$ and $Des(\mathbf{f}1)$. The blocks are now of two kinds, one for founder couples and one for founders which are not founder couples. The latter are treated as before. The founder couple blocks consist of the union of $Des(\mathbf{f}1)$, $Des(\mathbf{f}2)$ and $SecDes(\mathbf{f}1, \mathbf{f}1)$. The fixed bit from $SecDes(\mathbf{f}1, \mathbf{f}1)$ has to be placed first in the block. The algorithm now assumes, as before, knowledge about both fixed bits and the bits to be flipped and also about which bits to be swapped given by the function $swap$. We assume that the bits that are to be swapped two and two, are placed next to one another to make it simpler to implement the swapping in terms of MTBDDs. When the algorithm enters the founder couple block it reads the fixed bit from $SecDes(\mathbf{f}1, \mathbf{f}1)$, looks up the bits to be swapped and passes it on as a parameter and swaps them when it reaches them.

We will not give further details of the founder couple reduced version of the algorithm but refer to the implementation for the details.

Further Improvements: As we saw in the Lander-Green Algorithm in Algorithm 5, after having calculated the convolution $sp_{i-1}*t$, the result is multiplied with sp_i. Therefore in many cases, if sp_i is a sparse tree we can gain efficiency by giving (a BDD abstraction of) sp_i as a parameter in the algorithm for the convolution. Thus, if $sp_i(v) = 0$ we do not need to calculate the value for $sp_{i-1} * t(v)$. This feature is implemented in our algorithm.

Another improvement is obtained by delaying convolution of bits that are uninformative at the locus a distribution is being convolved to, and only calculating convolution probabilities that are needed at that locus.

7   Testing Results and Comparison with Existing Tools

Our implementation takes advantage of genetic information and is therefore able to handle almost arbitrarily large pedigrees if genetic information is close to being complete, but it is also a significant improvement when genetic information is not complete. As an example we take a real pedigree with 25 non-founding and 11 founding members (complexity level is 39bits) with data available for 19 micro satellite markers. Our implementation is able to analyze this pedigree in 5 minutes, while neither the current publicly available version of MERLIN (version 0.10.2) or GENEHUNTER (version 2.1-r5) are able to make any progress. By removing family members until complexity level of the pedigree is down to 25 bits MERLIN was able to analyze it in just over two minutes while GENEHUNTER needed 33 minutes, and our implementation took a second. If the complexity was moved back up to 27 bits MERLIN ran out of memory , while GENEHUNTER refused to perform the analysis because of foreseeable lack of memory, and our implementation took a second.

This new multi point linkage analysis algorithm has been implemented into a new version of linkage analysis package of DeCode Genetics, Allegro version 2, and will be made publicly available for both academic and commercial users. Those who are interested in the tool are welcome to contact Daniel Gudbjartsson at DeCode Genetics (email: dgf@decode.is).

References

1. L. Aceto, Jens A. Hansen, Anna Ingólfsdóttir, Jacob Johnsen and John Knudsen. The Complexity of Checking Consistency of Pedigree Information and Related Problems. Special issue on bioinformatics (Paola Bonazzoni, Gianluca Dalla Vedova and Tao Jiang guest editors) of the *Journal of Computer Science and Technology* 19(1):42–59, January 2004.
2. Abecasis GR, Cherny SS, Cookson WO, Cardon LR (2001) Merlin - rapid analysis of dense genetic maps using sparse gene flow trees. *Nature Genetics* 30:97–101

3. Andersen, H. R., An Introduction to Binary Decision Diagrams, Lecture notes, Technical University of Denmark, 1997

4. Bahar, R. I.,Frohm, E. A., Gaona, C. M., Hachtel, G. D., Macii, E.,Pardo, A., Somenzi, F. Algebraic Decision Diagrams and their Applications *Formal Methods in Systems Design*, 10(2/3):171–206, 1997

5. Bryant, R. E., Graph-based Algorithms for Boolean Function Manipulation. *IEEE Transactions on Computers*, 8(C-35):677-691, 1986

6. Clarke, E., Fujita, M., McGeer, P. Yang, J., Zhao, X. Multi-Terminal Binary Decision Diagrams: An Efficient Data Structure for Matrix Representation. International Workshop on Logic Synthesis, 1993.

7. Clarke, E. M., McMillan, K. L.,Zhao, X.,Fujita, M., Yan, J. C. Y., Spectral Transforms for Large Boolean Functions with Applications to Technology Mapping 30th ACM/IEEE DAC, Dallas, TX, 54–60, 1993.

8. Elston RC, Stewart J (1971) A general model for the genetic analysis of pedigree data. *Hum Hered* 21:523-542

9. Robert W. Cottingham Jr., Ramana M. Idury, and Alejandro A. Schaffer. Faster sequential genetic linkage computations. *American journal of human genetics*, 53:252–263, 1993.

10. Gudbjartsson, D. F, Jonasson, K, Frigge, M, Kong, A. (2000) Allegro, a new computer program for multipoint linkage analysis. *Nature Genetics* 25:12–13

11. Gudbjartsson, D., Hansen, J. A., Ingólfsdóttir, A., Johnsen, J, Knudsen, J., Single Point Algorithms in Genetic Linkage Analysis Comuter Aided Systems Theory-Eurocast 2003, 9th International Workshop on Computer Aided Systems Theory, Las Palmas de Gran Canaria, Spain, p.372–383, 2003,

12. Ingolfsdottir, A., Gudbjartsson, D., Gunnarsson, G., Thorvaldsson, Th., BDD-Based Algorithms in Genetic Linkage Analysis To appear in the proceedings of NETTAB 2004, Camerino, Italy, LNCS.

13. Idury, R. M, Elston, R, C., A Faster and More General Hidden Markov Model Algorithm for Multi Point Calculations. *Human Heredity*, 47:197–202, 1997.

14. Leonid Kruglyak, Mark J. Daly, Mary Pat Reeve-Daly, and Eric S. Lander. Parametric and Nonparametric Linkage Analysis: A Unified Multipoint Approach.

15. Leonid Kruglyak and Eric S. Lander. Faster multipoint linkage analysis using fourier transforms. *Journal of Computational Biology*, 5(1):7, 1998.

16. Lange, K. *Mathematical and Statistical Methods for Genetic Analysis*. Springer, 1997.

17. Eric S. Lander and Philip Green. Construction of Multilocus enetic Linkage Maps in Humans. *Proc. Natl. Acad. Sci.*, 84:2363–2367, 1987.

18. Kruglyak L, Daly MJ, Reeve-Daly MP, Lander ES (1996) Parametric and nonparametric linkage analysis: a unified multipoint approach. *Am J Hum Genet* 58:1347–1363

19. Kyriacos Markianos, Mark J. Daly, and Leonid Kruglyak. Efficient multipoint linkage analysis through reduction of inheritance space. *American journal of human genetics*, 68:963–977, 2001.

20. Jurg Ott. *Analysis of Human Genetic Linkage,Third Edition*. The Johns Hopkins University Press, 1999.

21. Somenzi, F., CUDD: CU decision diagram package. Public software, Colorado University, Boulder, 1997.

22. O'Connell JR. Rapid Multipoint Linkage Analysis via Inheritance Vectors in the Elston-Stewart Algorithm. *Hum Hered* 51:226-240, 2001.

23. [RJ86] Rabiner, L. R., Juang, B. H. An introduction to hidden Markov models. *IEEE ASSP Magazine*, 4-6, 1986.

Abstract Machines of Systems Biology

Luca Cardelli

Microsoft Research

Abstract. Living cells are extremely well-organized autonomous systems, consisting of discrete interacting components. Key to understanding and modeling their behavior is modeling their system organization. Four distinct chemical toolkits (classes of macromolecules) have been characterized, each combinatorial in nature. Each toolkit consists of a small number of simple components that are assembled (polymerized) into complex structures that interact in rich ways. Each toolkit abstracts away from chemistry; it embodies an abstract machine with its own instruction set and its own peculiar interaction model. These interaction models are highly effective, but are not ones commonly used in computing: proteins stick together, genes have fixed output, membranes carry activity on their surfaces. Biologists have invented a number of notations attempting to describe these abstract machines and the processes they implement. Moving up from molecular biology, *systems biology* aims to understand how these interaction models work, separately and together.

1 Introduction

Following the discovery of the structure of DNA, just over 50 years ago, molecular biologists have been unraveling the functioning of cellular components and networks. The amount of molecular-level knowledge accumulated so far is absolutely amazing. And yet we cannot say that we understand *how a cell works*, at least not to the extent of being able to easily modify or repair a cell. The process of understanding cellular components is far from finished, but it 4is becoming clear that simply obtaining a full part list will not tell us how a cell works. Rather, even for substructures that have been well characterized, there are significant difficulties in understanding how components interact as *systems* to produce the observed behaviors. Moreover, there are just too many components, and too few biologists, to analyze each component in depth in reasonable time. Similar problems occur also at each level of biological organization above the cellular level.

Enter *systems biology*, which has two aims. The first is to obtain massive amounts of information about whole biological systems, via high-throughput experiments that provide relatively shallow and noisy data. The Human Genome Project is a prototypical example: the knowledge it accumulated is highly valuable, and was obtained in an automated and relatively efficient way, but is just the beginning of understanding the human genome. Similar effort are now underway in *genomics* (finding the collection of all genes, for many genomes), in *transcriptomics* (the collection of all actively transcribed genes), in *proteomics* (the collection of all

C. Priami et al. (Eds.): Trans. on Comput. Syst. Biol. III, LNBI 3737, pp. 145–168, 2005.
© Springer-Verlag Berlin Heidelberg 2005

proteins), and in *metabolomics* (the collection of all metabolites). *Bioinformatics* is the rapidly growing discipline tasked with collecting and analyzing such *omics* data.

The other aim of syst4ems biology is to build, with such data, a science of the *principles of operation* of biological systems, based on the *interactions between components*. Biological systems are obviously well-engineered: they are very complex and yet highly structured and robust. They have only one major engineering defect: they have not been designed, in any standard sense, and so are not laid out as to be easily understood. It is not clear that any of the engineering principles of operations we are currently familiar with are fully applicable. Understanding such principles will require an interdisciplinary effort, using ideas from physics, mathematics, and computing. These, then, are the promises of systems biology: it will teach us new principles of operation, likely applicable to other sciences, and it will leverage other sciences to teach us *how cells work* in an actionable way.

In this paper, we look at the organization of biological systems from an information science point of view. The main reason is quite pragmatic: as we increasingly map out and understand the complex interactions of biological components, we need to *write down* such knowledge, in such a way that we can inspect it, animate it, and understand its principles. For genes, we can write down long but structurally simple strings of nucleotides in a 4-letter alphabet, that can be stored and queried. For proteins we can write down strings of amino acids in a 20-letter alphabet, plus three-dimensional information, which can be stored a queried with a little more difficulty. But how shall we write down *biological processes*, so that they can be stored and queried? It turns out that biologists have already developed a number of informal notation, which will be our starting points. These notations are abstractions over chemistry or, more precisely, are abstractions over a number of biologically relevant chemical toolkits.

2 Biochemical Toolkits

Apart from small molecules such as water and some metabolites, there are four large classes of *macromolecules* in a cell. Each class is formed by a small number of units that can be combined systematically to produce structures of great complexity. That is, to produce both individual molecules of essentially unbounded size, and multi-molecular complexes.

The four classes of macromolecules are as follows. Different members of each class can have different functions (structure, energy storage, etc.). We focus on the most combinatorial, information-bearing, members of each class:

- *Nucleic acids.* Five kinds of *nucleotides* combine in ordered sequences to form two nucleic acid polymers: *DNA* and *RNA*. As data structures, RNA is **lists**, and DNA is **doubly-linked lists**. Their most prominent role is in coding information, although they also have other important functions.
- *Proteins.* About 20 kinds of *amino acids* combine linearly to form proteins. Each protein folds in a specific three-dimensional shape (sometimes from multiple strings of amino acids). The main and most evolutionary stable property of a protein is not the exact sequence of amino acids that make it up, nor the exact folding process, but its collection of surface *features* that determine its function.

As data structures, proteins are **records** of features and, since these features are often active and stateful, they are **objects** in the object-oriented programming sense.

- *Lipids*: Among the lipids, *phospholipids* have a modular structure and can self-assemble into closed double-layered sheets (membranes). Membranes differ in the proportion and orientation of different phospholipids, and in the kinds of proteins that are attached to them. As data structures, membranes are **containers**, but with an active surface that acts as an **interface** to its contents.

- *Carbohydrates*: Among the carbohydrates, *oligosaccharides* are sugars linked in a branching structure. As data structures, oligosaccharides are **trees**. They have a vast number of configurations, and a complex assembly processes. *Polysaccharides* form even bigger structures, although usually of a semi-regular kind (rods, meshes). We do not consider carbohydrates further, although they are probably just as rich and interesting as the other toolkits. They largely have to do with energy storage and with cell surface and extracellular structures. But it should be noted that they too have a computational role, in forming unique surface structures that are subject to recognition. Many proteins are grafted with carbohydrates, through a complex assembly process called glycosylation.

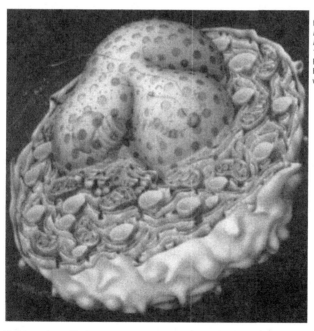

From *MOLECULAR CELL BIOLOGY, 4/e* by Harvey Lodish, *et. al.* ©1986, 1990, 1995, 2000 by W.H. Freeman and Company. Figure 1-1, Page 1. Used with permission.

Eukaryotic cells have an extensive array of membrane-bound compartments and organelles with up to 4 levels of nesting. The nucleus is a double membrane. The external membrane is less than 10% of the total.

Fig. 1. Eukaryotic Cell

Out of these four toolkits arises all the *organic chemicals*, composing, e.g., eukaryotic cells (Figure 1, [32] p.1). Each toolkit has specific structural properties (as emphasized by the bolded words above), systematic functions, and a peculiarly rich and flexible mode of operation. These peculiar modes of operation and systematic functions are what we want to emphasize, beyond their chemical realization.

Cells are without doubt, in many respects, information processing devices. Without properly processing information from their environment, they soon die for lack of nutrients or for predation. The blueprint of a cell, needed for its functioning and reproduction, is stored as digital information in the genome; an essential step of reproduction is the copying of that digital information. There are hints that information processing in the genome of higher organisms is much more sophisticated than currently generally believed [33].

We could say that cells are based on chemistry that also perform some information processing. But we take a more extreme position, namely that cells are chemistry *in the service* of information processing. Hence, we look for information processing machinery within the cellular machinery, and we try to understand the functioning of the cell in terms of information processing, instead of chemistry. In fact, we can readily find such information processing machinery in the chemical toolkits that we just described, and we can switch fairly smoothly from the classical description of cellular functioning in terms of classes of macromolecules, to a description based on abstract information-processing machines.

3 Abstract Machines

An *abstract machine* is a fictional information-processing device that can, in principle, have a number of different physical realizations (mechanical, electronic, biological, or even software). An abstract machine is characterized by:

- A collection of discrete states.
- A collection of operations (or events) that cause discrete transitions between states.

The evolution of states through transitions can in general happen concurrently. The adequacy of this generic model for describing complex systems is argued, e.g., in [22].

Each of the chemical toolkits we have just described can be seen as a separate abstract machine with an appropriate set of states and operations. This abstract interpretations of chemistry is by definition fictional, and we must be aware of its limitation. However, we must also be aware of the limitations of *not* abstracting, because then we are in general limited to work at the lowest level of reality (quantum mechanics) without any hope of understanding higher principles of organization. The abstract machines we consider are each grounded in a different chemical toolkit (nucleotides, amino acids, and phospholipids), and hence have some grounding in reality. Moreover, each abstract machine corresponds to a different kind of informal *algorithmic notation* that biologists have developed (Figure 2, bubbles): this is further evidence that abstract principles of organization are at work.

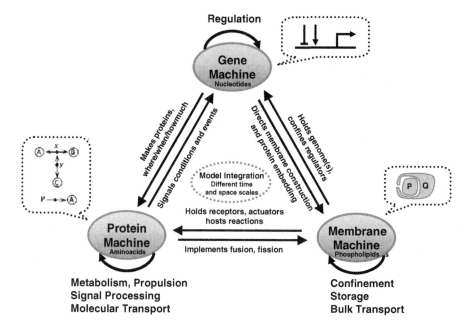

Fig. 2. Abstract Machines, Molecular Basis, and Notations

The *Gene Machine* (better known as Gene Regulatory Networks) performs information processing tasks within the cell. It regulates all other activities, including assembly and maintenance of the other machines, and the copying of itself. The *Protein Machine* (better known as Biochemical Networks) performs all mechanical and metabolic tasks, and also some signal processing. The *Membrane Machine* (better known as Transport Networks) separates different biochemical environments, and also operates dynamically to transport substances via complex, discrete, multi-step processes.

These three machines operate in concert and are highly interdependent. Genes instruct the production of proteins and membranes, and direct the embedding of proteins within membranes. Some proteins act as messengers between genes, and others perform various gating and signaling tasks when embedded in a membrane. Membranes confine cellular materials and bear proteins on their surfaces. In eukaryotes, membranes confine the genome, so that local conditions are suitable for regulation, and confine other reactions carried out by proteins in specialized vesicles.

Therefore, to understand the functioning of a cell, one must understand also how the various machines interact. This involves considerable difficulties (e.g. in simulations) because of the drastic difference in time and size scales: proteins interacts in tiny fractions of a second, while gene interactions take minutes; proteins are large molecules, but are dwarfed by chromosomes, and membranes are larger still. Before looking at the interactions among the different machine in more detail, we start by discussing each machine separately.

4 The Protein Machine (Biochemical Networks)

4.1 Principles of Operation

Proteins are long folded-up strings of amino acids with precisely determined, but often mechanically flexible, three-dimensional shapes. If two proteins have surface regions that are complementary (both in shape and in charge), they may stick to each other like Velcro, forming a protein **complex** where a multitude of small atomic forces crates a strong bond between individual proteins. They can similarly stick highly selectively to other substances. During a **complexation** event, a protein may be bent or opened, thereby revealing new interaction surfaces. Through complexation many proteins act as enzymes: they bring together compounds, including other proteins, and greatly facilitate chemical reactions between them without being themselves affected.

Proteins may also chemically modify each other by attaching or removing small phosphate groups at specific sites. Each such site acts as a boolean switch: over a dozen of them can be present on a single protein. Addition of a phosphate group (**phosphorilation**) is performed by an enzyme that is then called a **kinase**. Removal of a phosphate group (**dephosphorilation**) is performed by an enzyme that is then called a **phosphatase**. For example, a *protein phosphatase kinase kinase* is a protein that phosphorilates a protein that phosphorilates a protein that dephosphorilates a protein. Each (de-)phosphorilation may reveal new interaction surfaces, and each surface interaction may expose new phosphorilation sites.

It turns out that a large number of protein interactions work at the level of abstraction just described. That is, we can largely ignore chemistry and the protein folding process, and think of each protein as a collection of features (binding sites and phosphorilation sites) whose availability is affected by (de-)complexation and (de-)phosphorilation interactions. This abstraction level is emphasized in Kohn's Molecular Interaction Maps graphical notation [29][27] (Figure 4).

We can describe the operation of the *protein machine* as follows (Figure 3). Each protein is a collection of *sites* and *switches*; each of those can be, at any given time, either *available* or *unavailable*. Proteins can join at matching sites, to form bigger and bigger *complexes*. The availability of sites and switches in a complex is the *state* of the complex. A *system* is a multiset of (disjoint) complexes, each in a given state.

The protein machine has two kinds of operations. (1) An available switch on a complex can be turned on or off, resulting in a new state where a new collection of switches and sites is available. (2) Two protein complexes can combine at available sites, or one complex can split into two, resulting in a new state where a new collection of switches and sites is available.

Who is driving the switching and binding? Other proteins do. There are tens of thousands of proteins in a cell, so the protein machine has tens of thousands of+ "primitive instructions"; each with a specific way of acting on other proteins (or metabolites). For each cellular subsystem one must list the proteins involved, and how each protein interacts with the other proteins in terms of switching and binding.

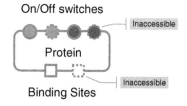

On/Off switches

Inaccessible

Protein

Inaccessible

Binding Sites

Each protein has a structure
of binary switches and binding sites.
But not all may always be *accessible*.

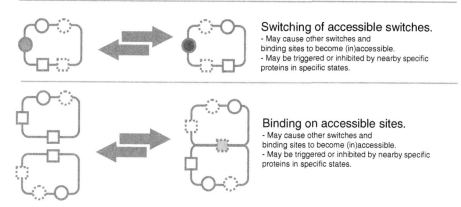

Switching of accessible switches.
- May cause other switches and
binding sites to become (in)accessible.
- May be triggered or inhibited by nearby specific
proteins in specific states.

Binding on accessible sites.
- May cause other switches and
binding sites to become (in)accessible.
- May be triggered or inhibited by nearby specific
proteins in specific states.

Fig. 3. The Protein Machine Instruction Set

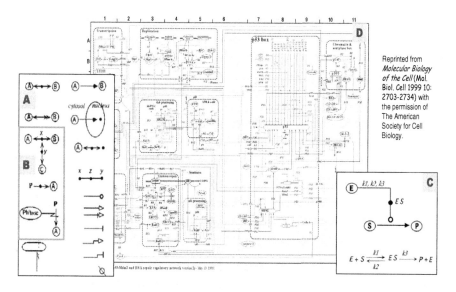

Reprinted from
*Molecular Biology
of the Cell* (Mol.
Biol. Cell 1999 10:
2703-2734) with
the permission of
The American
Society for Cell
Biology.

From [29]. **A**: graphical primitives. **B**: complexation and phosphorilation. **C**: enzymatic
diagram and equivalent chemical reactions. **D**: map of the p53-Mdm2 and DNA Repair
Regulatory Network.

Fig. 4. Molecular Interaction Maps Notation

4.2 Notations

Finding a suitable language in which to cast such an abstraction is a non-trivial task. Kohn designed a graphical notation, resulting in pictures such as Figure 4 [29]. This was a tremendous achievement, summarizing hundreds of technical papers in page-sized pictures, while providing a sophisticated and expressive notation that could be translated back into chemical equations according to semi-formal guidelines. Because of this intended chemical semantics, the dynamics of a systems is implied in Kohn's notation, but only by translation to chemical (and hence kinetic) equations. The notation itself has no dynamics, and this is one of its main limitation. The other major limitation is that, although graphically appealing, it tends to stop being useful when overflowing the borders of a page or of a whiteboard (the original Kohn maps span several pages).

Other notations for the protein machine can be devised. Kitano, for example, improved on the conciseness, expressiveness, and precision of Kohn's notation [28], but further sophistication in graphical notation is certainly required along the general principles of [18]. A different approach is to devise a textual notation, which inherently has no "page-size" limit and can better capture dynamics; examples are Bio-calculus [38], and most notably κ-calculus [14][15], whose dynamics is fully formalized. But one may not need to invent completely new formalisms. Regev and Shapiro, in pioneering work [49][47], described how to represent chemical and biochemical interactions within existing process calculi (π-calculus). Since process calculi have a well understood dynamics (better understood, in fact, than most textual notations that one may devise just for the purpose), that approach also provides a solid basis for studying systems expressed in such a notation. Finally, some notations incorporate both continuous and discrete aspects, as in Charon [3] and dL-systems [45].

4.3 Example: MAPK Cascade

The relatively simple Kohn map in 0 (adapted from [25]) describes the behavior of a circuit that causes Boolean-like switching of an output signal in presence of a very weak input signal. (It can also be described as a list of 10 chemical reactions, or of 25 differential/ algebraic equations, but then the network structure is not so apparent.) This network, generically called a MAPK cascade, has multiple biochemical implementations and variations. The components are proteins (enzymes, kinases, phophatases, and intermediaries). The circle-arrow Kohn symbol for "enzyme-assisted reaction" can signify here either a complexation that facilitates a reaction, or a phosphorilation/dephosphorilation, depending on the specific proteins.

The system initially contains reservoirs of chemicals KKK, KK, and K (say, 100 molecules each), which are transformed by the cascade into the kinases KKK*, KK-PP, and K-PP respectively. Enzymes E2, KK-Phosphatase and K-Phosphatase are always available (say, 1 molecule each), and tend to drive the reactions back. Appearance of the input enzyme E1 in very low numbers (say, less than 5) causes a

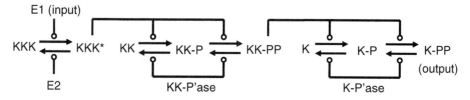

Fig. 5. ++ MAPK Cascade

sharp (Boolean-like 0-100) transition in the concentration of the output K-PP. The concentrations of the intermediaries KK-PP, and especially KKK*, raise in a much smoother, non-Boolean-like, fashion. Given the mentioned concentrations, the network works fine by setting all reaction rates to equal values.

To notice here is that the detailed description of each of the individual proteins, with their folding processes, surface structures, interaction rates under different conditions, etc. could take volumes. But what makes this signal processing network work is the structure of the network itself, and the relatively simple interactions between the components.

4.4 Summary

The fundamental flavor of the Protein Machine is: <u>fast synchronous binary interactions</u>. Binary because interactions occur between two complementary surfaces, and because the likelihood of three-party instantaneous chemical interactions can be ignored. Synchronous because both parties potentially feel the effect of the interaction, when it happens. Fast because individual chemical reactions happen at almost immeasurable speeds. The parameters affecting reaction speed, in a well-stirred solution, are just a reaction-specific rate constant having to do with surface affinity, plus the concentrations of the reagents (and the temperature of the solution, which is usually assumed constant). Concentration affects the likelihood of molecules randomly finding each other by Brownian motion. Note that Brownian motion is surprisingly effective at a cellular scale: a molecule can "scan" the equivalent volume of a bacteria for a match in 1/10 of a second, and it will in fact scan such a bounded volume because random paths in 3D do not return to the origin.

5 The Gene Machine (Gene Regulatory Networks)

5.1 Principles of Operation

The *central dogma of molecular biology* states that DNA is transcribed to RNA, and RNA is translated to proteins (and then proteins do all the work). This dogma no longer paints the full picture, which has become considerably more detailed in recent years. Without entering into a very complex topic [33], let us just note that some proteins go back and bind to DNA. Those proteins are called **transcription factors** (either **activators** or **repressors**); they are produced for the purpose of allowing one gene (or signaling pathway) to communicate with other genes. Transcription factors are not simple messages: they are proteins, which means they are subject to complexation, phosphorilation, and programmed degradation, which all have a role in gene regulation.

A **gene** is a stretch of DNA consisting of two (not necessarily contiguous or unbroken) regions: an *input (**regulatory**) region*, containing **protein binding sites** for transcription factors, and an *output (**coding**) region*, coding for one or more proteins that the gene produces. Sometimes there are two coding regions, in opposite directions [46], on count of DNA being a doubly-linked list. Sometimes two genes overlap on the same stretch of DNA.

The output region functions according to the **genetic code**: a well understood and almost universal table mapping triplets of nucleotides to one of about 20 amino acids, plus start and stop triplets. The input region functions according to a much more complex code that is still poorly understood: transcription factors, by their specific 3D shapes, bind to specific nucleotide sequences in the input region, with varying binding strength depending of the precision of the match.

Thus, the gene machine, although entirely determined by the digital information coded in DNA, is not entirely digital in functioning: a digitally encoded protein, translated and folded-up, uses its "analog" shape to recognize another digital string and promote the next step of translation. Nonetheless, it is customary to ignore the details of this process, and simply measure the effectiveness with which (the product of) a gene affects another gene. This point of view is reflected in standard notation for gene regulatory networks (Figure 7).

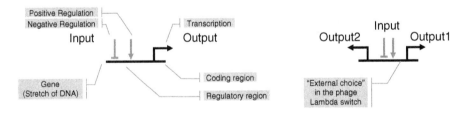

Fig. 6. The Gene Machine Instruction Set

In Figure 6, a gene is seen as a hardware gate, and the genome can be seen as a vast circuit composed of such gates. Once the performance characteristics of each gate is understood, one can understand or design circuits by combining gates, almost as one would design digital or analog hardware circuits. The performance characteristics of each gene in a genome is probably unique. Hence, as in the protein machine, we are going to have thousands of "primitive instructions": one for each gene.

A peculiarity of the gene machine is that a set of gates also determines the network connectivity. This is in contrast with a hardware circuit, where there is a collection of gates out of a very small set of "primitive gates", and then a separate wiring list. Each gene has a *fixed output*; the protein the gene codes for (although post-processing may vary such output). Similarly, a gene has a *fixed input*: the fixed set of binding sites in the input region. Therefore, by knowing the nucleotide sequence of each gene in a genome, one (in principle) also knows the network connectivity without further information. This situation is similar to a software assembly-language program: "Line 3: Goto Line 5" where both the input and output addresses are fixed, and the flow graph is determined by the instructions in the program. However, a further difference is that the output of a gene is not the "address" of another gene: it is a protein that can bind with varying strength to a number of other genes.

The state of a gene machine is the concentrations of the transcription factors produced by each gene (or arriving from the environment). The operations, again, are the input-output functions of each gene. But what is the "execution" of a gene machine? It is not as simple as saying that one gene stimulates or inhibits another gene. It is known that certain genes perform complex computations on their inputs that are a mixture of boolean, analog, and multi-stage operators (Figure 7-B [54]). Therefore, the input region of each gene can itself be a sophisticated machine.

Whether the execution of a gene machine should be seen as a continuous or discrete process, both in time and in concentration levels, is already a major question. Qualitative models (e.g.: random and probabilistic Boolean networks [26][50], asynchronous automata [52], network motifs [36]) can provide more insights that quantitative models, whose parameters are hard to come by and are possibly not critical. On the other hand, it is understood that pure Boolean models are inadequate in virtually all real situations. Continuous, stochastic, and decay aspect of transcription factor concentrations are all critical in certain situations [34][53].

5.2 Notations

Despite all these difficulties and uncertainties, a single notation for the gene machine is in common use, which is the gene network notation of Figure 7-A. There, the gates are connected by either "excitatory" (pointed arrow) or "inhibitory" (blunt arrow) links. What such relationships might mean is often left unspecified, except that, in a common model, a single constant weight is attached to each link.

Any serious publication would actually start from a set of ordinary differential equations relating concentrations of transcription factors, and use pictures such at Figure 7-A only for illustration, but this approach is only feasible for small networks. The best way to formalize the *notation* of gene regulatory networks is still subject to debate and many variations, but there is little doubt that formalizing such a notation will be essential to get a grasp on gene machines the size of genomes (the smallest of which, M.Genitalium, is on the order of 150 Kilobytes, and one closer to human cellular organization, Yeast, is 3 Megabytes).

5.3 Example: Repressilator

The circuit in Figure 8, artificially engineered in E.Coli bacteria [19], is a simple oscillator (given appropriate parameters). It is composed of three genes with single input that inhibit each other in turn. The circuit gets started by *constitutive transcription*: each

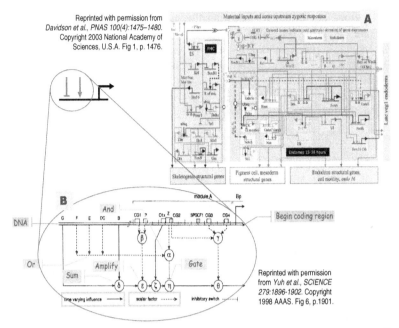

A [16]: gene regulatory network involved in sea urchin embryo development: **B** [54]: boolean/arithmetic diagram of module A, the last of 6 interlinked modules in the regulatory region of the *endo16* sea urchin gene; G,F,E,DC,B are module outputs feeding into A, the whole region is 2300 base pairs.

Fig. 7. Gene Regulatory Networks Notation

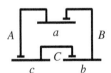

Fig. 8. Repressilator Circuit

gene autonomously produces output in absence of inhibition, and the produced output decays at a certain stochastic rate. The symmetry of the circuit is broken by the underlying stochastic behavior of chemical reactions. Its behavior con be understood as follows. Assume that gene *a* is at some point not inhibited (i.e. the product *B* of gene *b* is absent). Then gene *a* produces *A*, which shuts down gene *c*. Since gene *c* is no longer producing C, gene *b* eventually starts producing *B*, which shuts down gene *a*. And so on.

5.4 Summary

The fundamental flavor of the Gene Machine is: <u>slow asynchronous stochastic broadcast</u>. The interaction model is really quite strange, by computing standards. Each gene has a fixed output, which is not quite an address for another gene: it may bind to a large number of other genes, and to multiple locations on each gene. The transcription factor is produced in great quantities, usually with a well-specified time-to-live, and needs to reach a certain threshold to have an effect. On the other hand, various mechanisms can guarantee Boolean-like switching when the threshold is crossed, or, very importantly, when a message is *not* received. Activation of one gene by another gene is slow by any standard: typically one to five minutes, to build up the necessary concentration[1]. However, the genome can slowly direct the assembly-on-need of protein machines that then act fast: this "swap time" is seen in experiments that switch available nutrients. The stochastic aspect is fundamental because, e.g., with the same parameters, a circuit may oscillate under stochastic/discrete semantics, but not under deterministic/continuous semantics [53]. One reason is that a stochastic system may decay to *zero* molecules of a certain kind at a given time, and this can cause switching behavior, while a continuous system may asymptotically decay only to a non-zero level.

6   The Membrane Machine (Transport Networks)

6.1   Principles of Operation

A cellular membrane is an oriented closed surface that performs various molecular functions. Membranes are not just containers: they are coordinators and sites of major activity[2]. Large functional molecules (proteins) are embedded in membranes with consistent orientation, and can act on both sides of the membrane simultaneously. Freely floating molecules interact with membrane proteins, and can be sensed, manipulated, and pushed across by active molecular channels. Membranes come in different kinds, distinguished mostly by the proteins embedded in them, and typically consume energy to perform their functions. The consistent orientation of membrane proteins induces an orientation on the membrane.

One of the most remarkable properties of biological membranes is that they form a two-dimensional fluid (a lipid bilayer) embedded in a three-dimensional fluid (water). That is, both the structural components and the embedded proteins freely diffuse on the two-dimensional plane of the membrane (unless they are held together by specific mechanisms). Moreover, membranes float in water, which may contain other molecules that freely diffuse in that three-dimensional fluid. Membrane themselves are impermeable to most substances, such as water and protons, so that they partition the three-dimensional fluid. This organization provides a remarkable combination of freedom and structure.

Many membranes are highly dynamic: they constantly shift, merge, break apart, and are replenished. But the transformations that they support are naturally limited,

---

[1] Consider that bacteria replicate in only 20 minutes while cyclically activating hundreds of genes. It seems that, at lest for bacteria, the gene machine can make "wide" but not very "deep" computations [36].

[2] "For a cell to function properly, each of its numerous proteins must be localized to the correct cellular membrane or aqueous compartment." [32] p.675.

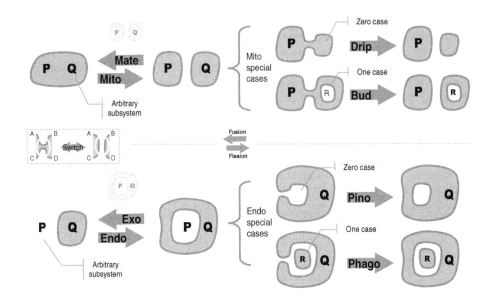

Fig. 9. The Membrane Machine Instruction Set (2D)

partially because membranes must preserve their proper orientation, and partially because membrane transformations need to be locally-initiated and continuous. For example, it is possible for a membrane to gradually buckle and create a bubble that then detaches, or for such a bubble to merge back with a membrane. But it is not possible for a bubble to "jump across" a membrane (only small molecules can do that), of for a membrane to turn itself inside-out.

The basic operations on membranes, implemented by a variety of molecular mechanisms, are *local fusion* (two patches merging) and *local fission* (one patch splitting in two) [8]. We discuss first the 2D case, which is instructive and for which there are some formal notations, and then the 3D case, the real one for which there are no formal notations.

In two dimensions (Figure 9), at the local scale of membrane patches, fusion and fission become indistinguishable as a single operation, *switch*, that takes two membrane patches, i.e. to segments A-B and C-D, and switches their connecting segments into A-C and B-D (crossing is not allowed). We may say that, in 2D, a switch is a fusion when it decreases the number of whole membranes, and is a fission when it increases such number.

When seen on the global scale of whole 2D membranes, switch induces four operations: in addition to the obvious splitting (Mito) and merging (Mate) of membranes, there are also operation, quite common in reality, that cause a membrane to "eat" (Endo) or "spit" (Exo) another subsystem (P). There are common special cases of Mito and Endo, when the subsystem P consists of zero (Drip, Pino) or one (Bud, Phago) membranes. All these operations *preserve bitonality* (dual coloring); that is, if a subsystem P is on a dark (or light) background before a reaction, it will be on a dark (or light) background after the reaction. Bitonality is related to preservation of membrane orientation, and to locality of operations (a membrane jumping across another one does not preserve bitonality). Bitonal operations ensure that what is or was *outside* the cell (light) never gets mixed with what is *inside* (dark). The main reactions that violate bitonality are destructive and non-local ones (such a digestion, not shown). Note that Mito/Mate preserve the nesting depth of subsystems, and hence they cannot encode Endo/Exo; instead, Endo/Exo can encode Mito/Mate [12].

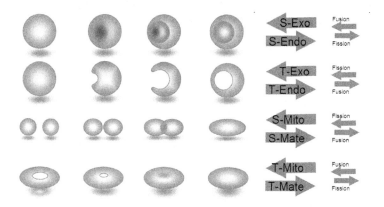

Each row consists of initial state, two intermediate states, and final state (and back).

Fig. 10. The Membrane Machine Instruction Set (3D)

In three dimensions, the situation is more complex (Figure 10). There are 2 distinct local operations on surface patches, inducing 8 distinct global operations that change surface topology. *Fusion* joins two Positively curved patches (in the shapes of domes) into one Negatively curved patch (in the shape of a hyperbolic cooling tower) by allowing the P-patches to kiss and merge. *Fission* instead splits one N-patch into two P-patches by pinching the N-patch. Fusion does not necessarily decrease the number of membranes in 3D (it may turn a sphere into a torus in two different ways: T-Endo T-Mito), and Fission does not necessarily increase the number of membranes (it may turn a torus into a sphere in two different ways: T-Exo, T-Mate). In addition, Fusion may merge two spheres into one sphere in two different ways (S-Exo, S-Mate), and Fission may split one sphere into two spheres in two different ways (S-Endo, S-Mito). Note that S-Endo and T-Endo have a common 2D cross section (Endo), and similarly for the other three pairs.

Cellular structures have very interesting dynamic topologies: the eukaryotic nuclear membrane, for example, is two nested spheres connected by multiple toroidal holes (and also connected externally to the Endoplasmic Reticulum). This whole structure is disassembled, duplicated, and reassembled during cellular mitosis. Developmental processes based on cellular differentiation are also within the realm of the Membrane Machine, although geometry, in addition to topology, is an important factor there.

6.2 Notations

The informal notation used to describe executions of the Membrane Machine does not really have a name, but can be seen in countless illustrations (e.g., Figure 11, [32] p.730). All the stages of a whole process are summarized in a single snapshot, with arrows denoting operations (Endo/Exo etc.) that cause transitions between states. This kind of depiction is natural because often all the stages of a process *are* observed at once, in photographs, and much of the investigation has to do with determining their proper sequence and underlying mechanisms. These pictures are usually drawn in two colors, which is a hint of the semantic invariant we call bitonality.

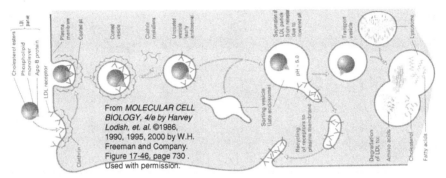

LDL particle (left) is recognized, ingested, and transported to a lysosome vesicle (right). [32], p.730.

Fig. 11. Transport Networks Notation

Some membrane-driven processes are semi-regular, and tend to return to something resembling a previous configuration, but they are also stochastic, so no static picture or finite-state-automata notation can tell the real story. Complex membrane dynamics can be found in the protein secretion pathway, through the Golgi system, and in many developmental processes. Here too there is a need for a precise dynamic notation that goes beyond static pictures; currently, there are only a few such notations [42][48][12].

6.3 Example: LDL Cholesterol Degradation

The membrane machine runs real algorithms: Figure 11 depicts LDL-cholesterol degradation. The "problem" this algorithm solves is to transport a large object (an LDL particle) to an interior compartment where it can be degraded; the particle is too big to just cross the membrane. The "solution", by a precise sequence of discrete steps and iterations, utilizes proteins embedded in the external cellular membrane and in the cytosol to recognize, bind, incorporate, and transport the particle inside vesicles to the desired compartment, all along recycling the active proteins.

6.4 Summary

The fundamental flavor of the Membrane Machine is: <u>fluid-in-fluid architecture, membranes with embedded active elements, and fusion and fission of compartments preserving</u> <u>*bitonality*</u>. Although dynamic compartments are common in computing, operations such as endocytosis and exocytosis have never explicitly been suggested as fundamental. They embody important invariants that help segregate cellular materials from environmental materials. The distinction between active elements *embedded* on the surface of a compartment, vs. active elements *contained* in the compartment, becomes crucial with operations such as Exo. In the former case, the active elements are retained, while in the latter case they are lost to the environment.

7 Three Machines, One System

7.1 Principles of Operation

We have discussed how three classes of chemicals, among others, are fundamental to cellular functioning: nucleotides (nucleic acids), amino acids (proteins), and phospholipids (membranes). Each of our abstract machines is based primarily on one of these classes of chemicals: amino acids for the protein machine, nucleotides for the gene machine, and phospholipids for the membrane machine.

These three classes of chemicals are however heavily interlinked and interdependent. The gene machine "executes" DNA to produce proteins, but some of those proteins, which have their own dynamics, are then used as control elements of DNA transcription. Membranes are fundamentally sheets of pure phospholipids, but in living cells they are heavily doped with embedded proteins which modulate membrane shape and function. Some protein translation happens only through membranes, with the RNA input on one side, and the protein output on the other side or threaded into the membrane.

Therefore, the abstract machines are interlinked as well, as illustrated in Figure 2. Ultimately, we will need a single notation in which to describe all three machines (and more), so that a whole organism can be described.

7.2 Notations

What could a single notation for all three machines (and more) look like? All formal notations known to computing, from Petri Nets to term-rewriting systems, have already been used to represent aspects of biological systems; we shall not even attempt a review here. But none, we claim, has shown the breadth of applicability and scalability of process calculi [35], partially because they are not a single notation, but a coherent conceptual framework in which one can derive suitable notations. There is also a general theory and notation for such calculi [37], which can be seen as the formal umbrella under which to unify different abstract machines.

Major progress in using process calculi for describing biological systems was achieved in Aviv Regev's Ph.D. thesis [47], where it is argued that one of the standard existing process calculi, π-calculus, enriched with a stochastic semantics [24][43][44], is extraordinarily suitable for describing both molecular-level interactions and higher levels of organization. The same stochastic calculus is now being used to describe genetic networks [30]. For membrane interactions, though, we need something beyond basic process calculi, which have no notion of compartments. Ambient Calculus [13] (which extends π-calculus with compartments) has been adapted [47][48] to represent biological compartments and complexes. A more recent attempt, Brane Calculus [12], embeds the biological invariants and 2D operations from Section 6.

These experiences point at process calculi as, at least, one of the most promising notational frameworks for unifying different aspects of biological representation. In addition, the process calculus framework is generally suitable for relating different levels of abstractions, which is going to be essential for feasibly representing biological systems of high architectural complexity.

Figure 12 gives a hint of the difference in notational approach between process calculi and more standard notations. Ordinary chemical reaction notation *is* a process calculus: it is a calculus of chemical processes. But it is a notation that focuses on reactions instead of components; this becomes a disadvantage when components have rich structure and a large state space (like proteins). In chemical notation one describes each state of a component as a different chemical species (Na, Na^+), leading to an combinatorial blowup in the *description* of the system (the blowup carries over to related descriptions in terms of differential equations). In process calculus notation, instead, the components are described separately, and the reactions (occurring through complementary event pairs such as !r and ?r) come from the interactions of the components. Interaction leads to a combinatorial blowup in the *dynamics* of interactions, but not in the description of the systems, just like in ordinary object-oriented programming.

On the left of Figure 12 we have a chemical description of a simple system of reactions, with a related (non-composition) Petri Nets description. On the right we have a process calculus description of the same system, with a related (compositional) description in terms of interacting automata (e.g., Statecharts [22] with sync pseudostates). Both kinds of descriptions can take into account stochastic reaction rates (k1,k2), and both can be mapped to the same stochastic model (Continuous-Time Markov Chains), but the descriptions themselves have different structural properties. From a simulation point of view, the left-hand-side approach leads to large sparse matrices of chemical species vs. chemical reactions, while the right-hand-side approach leads to large multisets of interacting objects.

7.3 Example: Viral Infection

The example in Figure 13 (adapted from [2], p.279) is the "algorithm" that a specific virus, the Semliki Forest virus, follows to replicate itself. It is a sequence of steps that involve the dynamic merging and splitting of compartments, the transport of materials, the operation of several proteins, and the interpretation of genetic information. The algorithm is informally described in English below. A concise description in Brane Calculus is presented in [12], which encodes the infection process at high granularity, but *in its entirety*, including the membrane, protein, and gene aspects.

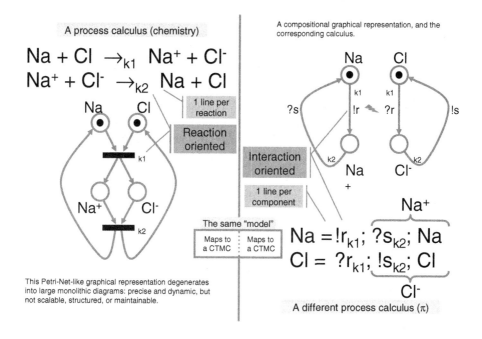

Fig. 12. Chemical vs. P+rocess Calculi Notations

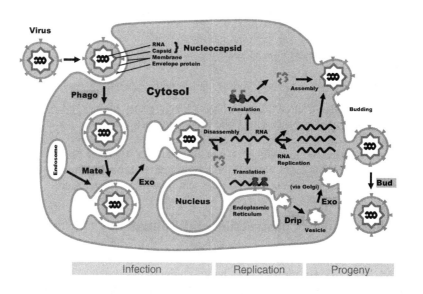

Fig. 13. Viral Replication

A virus is too big to cross a cellular membrane. It can either punch its RNA through the membrane or, as in this example, it can enter a cell by utilizing standard cellular phagocytosis machinery. The virus consists of a capsid containing the viral RNA (the nucleocapsid). The nucleocapsid is surrounded by a membrane that is similar to the cellular membrane (in fact, it is obtained from it "on the way out"). This membrane is however enriched with a special protein that plays a crucial trick on the cellular machinery, as we shall see shortly.

Infection: The virus is brought into the cell by phagocytosis, wrapped in an additional membrane layer; this is part of a standard transport pathway into the cell. As part of that pathway, an endosome merges with the wrapped-up virus. At this point, usually, the endosome causes some reaction to happen in the material brought into the cell. In this case, though, the virus uses its special membrane protein to trigger an exocytosis step that deposits the naked nucleocapsid into the cytosol. The careful separation of internal and external substances that the cell usually maintains has now been subverted.

Replication: The nucleocapsid is now in direct contact with the inner workings of the cell, and can begin doing damage. First, the nucleocapsid disassembles, depositing the viral RNA into the cytosol. This vRNA then follows three distinct paths. First it is replicated to provide the vRNA for more copies of the virus. The vRNA is also translated into proteins, by standard cellular machinery. The proteins forming the capsid are synthesized in the cytosol. The virus envelope protein is instead synthesized in the Endoplasmic Reticulum, and through various steps (through the Golgi apparatus) ends up lining transport vesicles that merge with the cellular membrane, along another standard transport pathway.

Progeny: In the cytosol, the capsid proteins self-assemble and incorporate copies of the vRNA to form new nucleocapsids. The newly assembled nucleocapsids make contact with sections of the cellular membrane that are now lined with the viral envelope protein, and bud out to recreate the initial virus structure outside the cell.

7.4 Summary

The fundamental flavor of cellular machinery is: chemistry in the service of materials, energy, and information processing. The processing of energy and materials (e.g., in metabolic pathways) need not be emphasized here, rather we emphasize the processing of information, which is equally vital for survival and evolution [1]. Information processing tasks are distributed through a number of interacting abstract machines with wildly different architectures and principles of operation.

8 Outlook: Model Construction and Validation

The biological systems we need to describe are massively concurrent, heterogeneous, and asynchronous: notoriously the hardest kinds of systems to cope with in programming. They have stochastic behavior and high resilience to drastic changes of environmental conditions. What organizational principles make these systems work reliably, and what conditions make them fail? These are the questions that computational modeling needs to answer.

There are two main aspects to modeling biological systems. *Model construction*, requires first an understanding of the principles of operation. This is what we have largely been discussing here: understanding the abstract machines of systems biology should lead us to formal notations that can be used to build (large, complex) biological models. But then there is *model validation*: a good scientific model has to be verified or falsified through postdiction and prediction. We briefly list different techniques that are useful for model validation, once a specific model has been written up in a specific precise notation.

Stochastic simulation of biochemical systems is a common technique, typically based on the physically well-characterized Gillespie algorithm [21], which originally was devised for reaction-oriented descriptions. The same algorithm can be used also for component-oriented (compositional) descriptions with a dynamically unbounded set of chemical species [44]. Stochastic simulation is particularly effective for systems with a relatively low number of interactions of any given kind, as is frequently the case in cellular-scale systems. It produces a single (high-likelihood) trace of the system for each run. It frequently reveals behavior that is difficult to anticipate, and that may not even correspond to continuous deterministic approximations [34]. It can be quantitatively compared with experiments.

Static analysis techniques of the kind common in programming can be applied to the description of biological systems [40]. Control-flow analysis and mobility analysis can reveal subsystems that *cannot* interact [7][41]. Causality analysis can reconstruct the familiar network diagrams from process description [11]. Abstract interpretation can be used to study specific facets of a complex model [39], including probabilistic aspects [17].

Modelchecking is now used routinely in the analysis of hardware and software systems that have huge state spaces; it is based on the state and transition model we emphasized during the discussion of abstract machines. Modelchecking consists of a model description language for building models, a query language for asking questions about models (typically temporal logic), and an efficient state exploration engine. The basic technology is very advanced, and is beginning to be applied to descriptions of biological systems too, in various flavors. **Temporal** modelchecking asks qualitative questions such as whether the systems can reach a certain state (and how), or whether a state is a necessary checkpoint for reaching another state [9][20]. **Quantitative** modelchecking asks quantitative questions about, e.g., whether a certain concentration can eventually equal or double some other concentration in some state [4][6]. **Stochastic** modelchecking, based, e.g., on discrete or continuous-time Markov chain models, can ask questions about the probability of reaching a given state [31].

Formal reasoning is the most powerful and hardest technique to use, but already there is a long tradition of building tools for verifying properties of concurrent systems. Typical activities in this area are checking behavioral equivalence between different systems, or between different abstraction levels of the same system, including now biological systems [10][5].

While computational approaches to biology and other sciences are now common, several of the techniques outlined above are unique to computer science and virtually unknown in other fields; hopefully they will bring useful tools and perspectives to biology.

9 Conclusions

Many aspects of biological organization are more akin to discrete hardware and software systems than to continuous systems, both in hierarchical complexity and in algorithmic-like information-driven behavior. These aspects need to be reflected in the modeling approaches and in the notations used to describe such systems, in order to make sense of the rapidly accumulating experimental data.

> *"The data are accumulating and the computers are humming, what we are lacking are the words, the grammar and the syntax of a new language..."*
>
> Dennis Bray (TIBS 22(9):325-326, 1997)

> *"The most advanced tools for computer process description seem to be also the best tools for the description of biomolecular systems."*
>
> Ehud Shapiro (Biomolecular Processes as Concurrent Computation, Lecture Notes, 2001)

> *"Although the road ahead is long and winding, it leads to a future where biology and medicine are transformed into precision engineering."*
>
> Hiroaki Kitano (Nature 420:206-210, 2002)

> *"The problem of biology is not to stand aghast at the complexity but to conquer it."*
>
> Sydney Brenner (Interview, Discover Vol. 25 No. 04, April 2004)

References

[1] C.Adami. What is complexity? BioEssays 24:1085–1094, Wiley, 2002.

[2] B.Alberts, D.Bray, J.Lewis, M.Raff, K.Roberts, J.D.Watson. Molecular biology of the cell. Third Edition, Garland.

[3] R.Alur, C.Belta, F.Ivancic, V.Kumar, M.Mintz, G.J.Pappas, H.Rubin, and J.Schug,. Hybrid modeling of biomolecular networks. Proceedings of the 4th International Workshop on Hybrid Systems: Computation and Control, Rome Italy, March 28-30, 2001. LNCS 2034.

[4] M.Antoniotti, B.Mishra, F.Park, A.Policriti, N.Ugel. Foundations of a query and simulation system for the modeling of biochemical and biological processes. In L.Hunter, T.A.Jung, R.B.Altman, A.K.Dunker, T.E.Klein, editors, The Pacific Symposium on Biocomputing (PSB 2003) 116-127. World Scientific, 2003.

[5] M.Antoniotti, C.Piazza, A.Policriti, M.Simeoni, B.Mishra. Modeling cellular behavior with hybrid automata: bisimulation and collapsing. In Int. Workshop on Computational Methods in Systems Biology (CMSB'03), LNCS. Springer, 2003. To appear.

[6] M.Antoniotti, A.Policriti, N.Ugel, B.Mishra. Model building and model checking for biochemical processes. In Cell Biochemistry and Biophysics, 2003. In press.

[7] C.Bodei, P.Degano, F.Nielson, H.R.Nielson. Control flow analysis for the pi-calculus. Proc. 9th International Conference on Concurrency Theory, LNCS 1466:84-98. Springer, 1998.

[8] K.N.J. Burger. Greasing membrane fusion and fission machineries. Traffic 1: 605–613. 2000.

[9] G.Ciobanu, V.Ciubotariu, B.Tanasa. A π-calculus model of the Na pump. Genome Informatics 13:469-471, 2002.

[10] G.Ciobanu. Software verification of biomolecular systems. In G.Ciobanu, G.Rozenberg (Eds.): Modelling in Molecular Biology, Natural Computing Series, Springer, 40-59, 2004.

[11] M.Curti, P.Degano, C.Priami, C.T.Baldari. Modelling biochemical pathways through enhanced pi-calculus. Theoretical Computer Science 325(1):111-140.

[12] L.Cardelli. Brane calculi – Interactions of biological membranes. Proc. Computational Methods in Systems Biology 2004. Springer. To appear.

[13] L.Cardelli, A.D.Gordon. Mobile ambients. Theoretical Computer Science, Special Issue on Coordination, D. Le Métayer Editor. Vol 240/1, June 2000. pp 177-213.

[14] V.Danos, M.Chiaverini. A core modeling language for the working molecular biologist. 2002.

[15] V.Danos, C.Laneve. Formal molecular biology. Theoretical Computer Science, to Appear.

[16] E.H.Davidson, D.R.McClay, L.Hood. Regulatory gene networks and the properties of the developmental process. PNAS 100(4):1475–1480, 2003.

[17] A.Di Pierro, H.Wiklicky. Probabilistic abstract interpretation and statistical testing. Proc. Second Joint International Workshop on Process Algebra and Probabilistic Methods, Performance Modeling and Verification. LNCS 2399:211-212. Springer, 2002.

[18] S.Efroni, D.Harel and I.R.Cohen. Reactive animation: realistic modeling of complex dynamic systems. IEEE Computer, to appear, 2005.

[19] M.B.Elowitz, S.Leibler. A synthetic oscillatory network of transcriptional regulators. Nature 403:335-338, 2000.

[20] F.Fages, S.Soliman, N.Chabrier-Rivier. Modelling and querying interaction networks in the biochemical abstract machine BIOCHAM. J. Biological Physics and Chemistry 4(2):64-73, 2004.

[21] D.T.Gillespie. Exact stochastic simulation of coupled chemical reactions, Journal of Physical Chemistry 81:2340–2361. 1977.

[22] D.Harel. Statecharts: a visual formalism for complex systems. Science of Computer Programming 8:231-274. North-Holland 1987.

[23] L.H.Hartwell, J.J.Hopfield , S.Leibler , A.W.Murray. From molecular to modular cell biology. Nature. 1999 Dec 2;402(6761 Suppl):C47-52.

[24] J. Hillston. A compositional approach to performance modelling. Cambridge University Press, 1996.

[25] C-Y.F.Huang, J.E.Ferrell Jr. Ultrasensitivity in the mitogen-activated protein kinase cascade. PNAS 93:10078–10083, 1996.

[26] S.Kauffman, C.Peterson, B.Samuelsson, C.Troein. Random Boolean network models and the yeast transcriptional network. PNAS 100(25):14796-14799, 2003.

[27] H.Kitano. The standard graphical notation for biological networks. The Sixth Workshop on Software Platforms for Systems Biology, 2002.

[28] H.Kitano. A graphical notation for biochemical networks. BIOSILICO 1:169-176, 2003.

[29] K.W.Kohn. Molecular interaction map of the mammalian cell cycle control and DNA repair systems. Molecular Biology of the Cell 10(8):2703-34, 1999.

[30] C.Kuttler, J.Niehren, R.Blossey. Gene regulation in the pi calculus: simulating cooperativity at the Lambda switch. BioConcur 2004, ENTCS.

[31] M.Kwiatkowska, G.Norman, D.Parker. Probabilistic symbolic model checking with PRISM: a hybrid approach. J. Software Tools for Technology Transfer (STTT), 6(2):128-142. Springer-Verlag, 2004.

[32] H.Lodish, A.Berk, S.L.Zipursky, P.Matsudaira, D.Baltimore, J.Darnell. Molecular cell biology. Fourth Edition, Freeman, 2002.

[33] J.S.Mattick. The hidden genetic program of complex organisms. Scientific American p.31-37, October 2004.

[34] H.H.McAdams, A.Arkin. It's a noisy business! Genetic regulation at the nanomolar scale. Trends Genet. 1999 Feb;15(2):65-9.

[35] R.Milner. Communicating and mobile systems: the π-calculus. Cambridge University Press, 1999.

[36] R.Milo, S.Shen-Orr, S.Itzkovitz, N.Kashtan, D.Chklovskii, U.Alon. Network motifs: simple building blocks of complex networks. Science 298:824-827, 2002.

[37] R.Milner. Bigraphical reactive systems. CONCUR 2001, Proc. 12th International Conference in Concurrency Theory, LNCS 2154:16-35, 2001.

[38] M.Nagasaki, S.Onami, S.Miyano, H.Kitano: Bio-calculus: its concept and molecular interaction. Genome Informatics 10:133-143, 1999. PMID: 11072350.

[39] F.Nielson, R.R.Hansen , H.R.Nielson. Abstract interpretation of mobile ambients. Science of Computer Programming, 47(2-3):145-175, 2003.

[40] F.Nielson, H.R.Nielson, C.Priami, D.Rosa. Static analysis for systems biology. Proc. ACM Winter International Symposium on Information and Communication Technologies. Cancun 2004.

[41] F.Nielson, H.R.Nielson, C.Priami, D.Rosa. Control flow analysis for BioAmbients. Proc. BioCONCUR 2003, to appear.

[42] G.Paun. Membrane computing. Springer, 2002.

[43] C.Priami. The stochastic pi-calculus. The Computer Journal 38: 578-589, 1995.

[44] C.Priami, A.Regev, E.Shapiro, W.Silverman. Application of a stochastic name-passing calculus to representation and simulation of molecular processes. Information Processing Letters, 80:25-31, 2001.

[45] P.Prusinkiewicz, M.Hammel, E.Mjolsness. Animation of plant development. Proceeding of SIGGRAPH 93. ACM Press, 351:360, 1993.

[46] M.Ptashne. Genetic switch: phage Lambda revisited. Cold Spring Harbor Laboratory Press. 3rd edition, 2004.

[47] A.Regev. Computational systems biology: a calculus for biomolecular knowledge. Ph.D. Thesis, Tel Aviv University, 2002.

[48] A.Regev, E.M.Panina, W.Silverman, L.Cardelli, E.Shapiro. BioAmbients: an abstraction for biological compartments. Theoretical Computer Science, to Appear.

[49] A.Regev, E.Shapiro: Cells as computation. Nature, 419:343, 2002.

[50] I.Shmulevich, E.R.Dougherty, W.Zhang. From Boolean to probabilistic Boolean networks as models of genetic regulatory networks. Proceedings of the IEEE 90(11):1778-1792, 2002.

[51] Systems biology markup language. http://www.sbml.org.

[52] D.Thieffry, R.Thomas. Qualitative analysis of gene networks. Pacific Symposium on Biocomputing 1998:77-88. PMID: 9697173.

[53] J.M.Vilar, H.Y.Kueh, N.Barkai, S.Leibler. Mechanisms of noise-resistance in genetic oscillators. PNAS, 99(9):5988–5992, 2002.

[54] C-H.Yuh, H.Bolouri, E.H.Davidson. Genomic cis-regulatory logic: experimental and computational analysis of a sea urchin gene. Science 279:1896-1902, 1998. www.sciencemag.org.

Author Index

Lecture Notes in Bioinformatics

Vol. 3745: J.L. Oliveira, V. Maojo, F. Martín-Sánchez, A.S. Pereira (Eds.), Biological and Medical Data Analysis. XII, 422 pages. 2005.

Vol. 3737: C. Priami, E. Merelli, P. P. Gonzalez, A. Omicini (Eds.), Transactions on Computational Systems Biology III. VII, 169 pages. 2005.

Vol. 3695: M.R. Berthold, R.C. Glen, K. Diederichs, O. Kohlbacher, I. Fischer (Eds.), Computational Life Sciences. XI, 277 pages. 2005.

Vol. 3692: R. Casadio, G. Myers (Eds.), Algorithms in Bioinformatics. X, 436 pages. 2005.

Vol. 3680: C. Priami, A. Zelikovsky (Eds.), Transactions on Computational Systems Biology II. IX, 153 pages. 2005.

Vol. 3678: A. McLysaght, D.H. Huson (Eds.), Comparative Genomics. VIII, 167 pages. 2005.

Vol. 3615: B. Ludäscher, L. Raschid (Eds.), Data Integration in the Life Sciences. XII, 344 pages. 2005.

Vol. 3594: J.C. Setubal, S. Verjovski-Almeida (Eds.), Advances in Bioinformatics and Computational Biology. XIV, 258 pages. 2005.

Vol. 3500: S. Miyano, J. Mesirov, S. Kasif, S. Istrail, P.A. Pevzner, M. Waterman (Eds.), Research in Computational Molecular Biology. XVII, 632 pages. 2005.

Vol. 3388: J. Lagergren (Ed.), Comparative Genomics. VII, 133 pages. 2005.

Vol. 3380: C. Priami (Ed.), Transactions on Computational Systems Biology I. IX, 111 pages. 2005.

Vol. 3370: A. Konagaya, K. Satou (Eds.), Grid Computing in Life Science. X, 188 pages. 2005.

Vol. 3318: E. Eskin, C. Workman (Eds.), Regulatory Genomics. VII, 115 pages. 2005.

Vol. 3240: I. Jonassen, J. Kim (Eds.), Algorithms in Bioinformatics. IX, 476 pages. 2004.

Vol. 3082: V. Danos, V. Schachter (Eds.), Computational Methods in Systems Biology. IX, 280 pages. 2005.

Vol. 2994: E. Rahm (Ed.), Data Integration in the Life Sciences. X, 221 pages. 2004.

Vol. 2983: S. Istrail, M.S. Waterman, A. Clark (Eds.), Computational Methods for SNPs and Haplotype Inference. IX, 153 pages. 2004.

Vol. 2812: G. Benson, R.D. M. Page (Eds.), Algorithms in Bioinformatics. X, 528 pages. 2003.

Vol. 2666: C. Guerra, S. Istrail (Eds.), Mathematical Methods for Protein Structure Analysis and Design. XI, 157 pages. 2003.